T0406501

Current Topics in Microbiology and Immunology

Volume 332

Alexander V. Karasev
Editor

Plant-produced Microbial Vaccines

Editor
Alexander V. Karasev
University of Idaho
Department of Plant, Soil & Entomological Sciences
Moscow ID 83844-2339
USA
akarasev@uidaho.edu

ISBN 978-3-540-70857-5 e-ISBN 978-3-540-70868-1
DOI 10.1007/978-3-540-70868-1

Current Topics in Microbiology and Immunology ISSN 0070-217x

Library of Congress Catalog Number: 2008936144

Cover design: WMX Design Gmbh, Heidelberg

Printed on acid-free paper

9 8 7 6 5 4 3 2 1

springer.com

Preface

In recent years, plants have been increasingly explored for the production of biomedicines and vaccine components. The two main advantages of plant systems are low cost and a greater potential for scalability as compared to microbial or animal systems. An additional advantage from the public health point of view is the high safety compared to animal systems, which is important for vaccine production: there are no known plant pathogens capable of replicating in animals and in humans, in particular. A particular antigen or a protein has to be expressed in a plant using one of the many available platforms; this antigen/protein subsequently needs to be purified or processed, and later formulated into a vaccine or a therapeutic; these need to be delivered to a human or animal body via an appropriate route. Naturally, all these vaccines and therapeutics must be subjected to regulatory approvals prior to their use. Thus, the challenge is to adapt plant-based platforms for the production of cost-efficient biomedicals that can be approved by FDA for use as vaccine components or therapeutics, which will be competitive against existing vaccines and drugs.

This volume attempts to address the entire spectrum of challenges facing the nascent field of plant-based biomedicals, from the selection of an appropriate production platform to specific methods of downstream processing and regulatory approval issues. The chapter by D.C. Hooper is devoted to immunological issues that can arise for antigens produced in plants and delivered to a human or an animal via different routes. This chapter also discusses such specific topics as tolerance and immunomodulation, with particular reference to oral delivery of plant produced antigens. The chapter by Smith et al. discusses one specific example of a virus-based platform for the expression of peptides in plants, and related issues of downstream processing, that is, manufacture and purification of virus particle-based vaccines, and final product release and stability. Another production platform, via chloroplast-based expression of proteins, is discussed in the chapter by S. Chebolu and H. Daniell. The chapter presents several examples of various vaccine components and other biomedicals produced in plants, and these range from bacterial and viral proteins to human serum proteins and antibodies. The chapter by Ko et al. describes a particular application of the plant-based production of human antibodies used for passive immunization against rabies. It is an example of a classical transgenic technology modified for a specific expression of two antibody chains. The chapter by R. Hammond and L. Nemchinov reviews the current status of plant production of veterinary vaccines, utilizing a great variety of platforms. The final chapter by C. Tacket presents case studies for the human trials of the first plant-produced candidate vaccines and discusses several regulatory issues that need to be addressed prior to their approval.

Production of vaccine components and other biomedicals in plants has a great potential in medicine and veterinary science. We hope that this volume will be a valuable contribution to this rapidly growing research field.

USA A.V. Karasev

Contents

Contributors

Robert Brodzik
Biotechnology Foundation Labs, Thomas Jefferson University, Philadelphia,
PA 19107, USA

S. Chebolu
University of Central Florida, Department of Molecular Biology & Microbiology,
336 Biomolecular Science (Bldg # 20), Orlando FL 32816-2360, USA
daniell@mail.ucf.edu

Henry Daniell
University of Central Florida, Department of Molecular Biology & Microbiology,
336 Biomolecular Science (Bldg # 20), Orlando FL 32816-2360, USA

Wayne P. Fitzmaurice
Large Scale Biology Corporation, Vacaville, CA 95688, USA

Rosemarie W. Hammond
USDA-ARS, BARC-West, Rm.252, Bldg. 011, Beltsville, MD20705, USA

Craig Hooper
Associate Professor, Center for Neurovirology, Kimmel Cancer Center,
Thomas Jefferson University, Philadelphia, PA, USA
Douglas.hooper@jefferson.edu

Alexander V. Karasev
University of Idaho, Department of PSES, Moscow, ID 83844-2339, USA
akarasev@uidaho.edu

Kisung Ko
Department of Biologcal Science, College of Natural Sciences, Wonkwang
University, Iksan, Jeonbuk 570-749, Korea
kko@mail.jci.tju.edu

Lev G. Nemchinov
USDA-ARS, BARC-West, Rm.252, Bldg. 011, Beltsville, MD20705, USA
Rose.hammond@ars.usda.gov

Kenneth E. Palmer
Owensboro Cancer Research Program, University of Louisville,
1020 Breckenridge Street, Owensboro, KY 42303, USA
Kenneth.palmer@louisville.edu

Mark L. Smith
Large Scale Biology Corporation, Vacaville, CA 95688, USA

Zeonon Steplewski
Biotechnology Foundation Labs, Thomas Jefferson University, Philadelphia,
PA 19107, USA

Carol O. Tacket
Center for Vaccine Development, University of Maryland School of Medicine,
685 West Baltimore St., Baltimore, MD 21201-1192
ctacket@medicine.umaryland.edu

Thomas H. Turpen
Technology Innovation Group, Inc., 7006 Firewheel Hollow, Austin,
TX 78750, USA
tturpen@techingroup.com

Plant Vaccines: An Immunological Perspective

D.C. Hooper

Contents

Abstract The advent of technologies to express heterologous proteins *in planta* has led to the proposition that plants may be engineered to be safe, inexpensive vehicles for the production of vaccines and possibly even vectors for their delivery. The immunogenicity of a variety of antigens of relevance to vaccination expressed in different plants has been assessed. The purpose of this article is to examine the utility of plant-expression systems in vaccine development from an immunological perspective.

D.C. Hooper(✉)
Center for Neurovirology, Kimmel Cancer Center, Thomas Jefferson University, Philadelphia, PA 19107-6731, USA
e-mail: douglas.hooper@jefferson.edu

A.V. Karasev (ed.) *Plant-produced Microbial Vaccines.*
Current Topics in Microbiology and Immunology 332

Introduction

From its genesis, vaccine development has primarily been an empirical science; originating with the chance discovery of Jenner that a relatively benign cowpox infection protected an individual against the considerably more deadly smallpox (Jenner 1798). Based on this observation and his own work on the germ theory of disease, Louis Pasteur pioneered the classic approach to vaccine development: attenuating the pathogenic agent such that its capacity to cause disease was limited but its antigenic structure unchanged (Pasteur 2002). During the heyday of vaccine development that followed, it was recognized that certain diseases caused by proteins, elaborated by bacteria and toxins, could be rendered apathogenic and used to vaccinate against the disease (Ramon and Zoeller 1927). Thus, by the beginning of the twentieth century, the basic attributes of a successful vaccine had been established. Essentially, the causative agent of a disease was modified to dissociate its ability to induce a protective immune response from its pathogenicity. Many of the vaccines for infectious diseases commonly in use today still consist of preparations of attenuated viruses or inactivated viruses and toxins (CDC 2002). The use of such reagents is termed active vaccination because their success is dependent upon the induction of an immune response in the recipient. This contrasts with passive immunization where the administration of preformed antibodies is used to confer immunity, as naturally occurs between the mother and offspring during pregnancy and early development.

Vaccination by infection with an attenuated variant of a pathogen should provide the best long-term protection by eliciting the full range of immune effectors and immunological memory, the capacity to rapidly mount a recall response to an antigen. Nevertheless, even a vaccine that induces an incomplete immune response may be considered successful if it prevents disease, the primary objective of vaccination. For example, passive vaccination can be very effective at neutralizing a disease-causing toxin but does not induce immunological memory. Nevertheless, in rabies postexposure prophylaxis the passive administration of virus-neutralizing antibodies provides an underlying active immune response the additional time required to develop and clear the infection so that rabies with its lethal outcome is avoided (Hanlon et al. 2001). Similarly, active vaccination with a noninfectious vaccine is unlikely to generate a cytotoxic CD8 T cell response, which generally requires infection of target cells, but can be efficacious in protecting against an intracellular pathogen. In this case, the response would limit the capacity of an infectious agent to invade (antibody) as well as provide the immunological memory (T helper cells) that would accelerate the induction of the CD8 T cells or other cytotoxic effectors required for clearance. Under normal circumstances, vaccination is not expected to prevent subsequent infection with a pathogen, as this would require the maintenance of high levels of pathogen-neutralizing antibodies at the point of entry. While this is theoretically possible through IgA secretion in the gut, in practice even the clearance of an enteric virus does not prevent subclinical infection with the same virus (Weinstein and Cebra 1991). Thus the objective of vaccination is to elicit an antibody response that interferes with the invasion of the target pathogen and to prime T helper cells such that the generation of a complete response encompassing humoral and cellular immunity is enhanced during the natural infection.

As understanding of immune function as well as the pathogenesis of different diseases has advanced, the uses of vaccination have expanded into areas distinct from protecting against infection. One of these areas is immunomodulation. An example of this would be passive immunization with Rh-specific antibodies to prevent sensitization of Rh-negative mothers against the Rh antigen, which has been used to modulate immunity for nearly 50 years (Kumpel 2002). Recently trials have been conducted to determine if active immunization can be used to treat autoimmune disease (Vandenbark et al. 2001; Cohen-Kaminsky and Jambou 2005; Li et al. 2005). Another rapidly growing use for vaccination is in cancer therapy. With the identification of antigens expressed exclusively or at higher levels than normal by transformed cells, a variety of active and passive anti-cancer vaccines are in development (Dredge et al. 2002; Vichier-Guerre et al. 2003; Lenarczyk et al. 2004; Bodey et al. 2000; Ko et al. 2005).

Molecular technologies enabling both antigens and antibodies to be expressed by plant-based systems have been developed. The purpose of this article is to examine the potential advantages and disadvantages of the *in planta* production of vaccine reagents from an immunological perspective.

Basic Immunology: Antigens and Immunogenicity

Antigens are the structures that are recognized by T cells, B cells and antibody in an immune response. The capacity of an antigen to induce an immune response is its immunogenicity. Most antigens are protein and not inherently immunogenic. This is likely to be at least partly due to the nature of antigen recognition. While B cells and antibodies can passively interact with intact antigens, T cells recognize antigens that have been processed and presented at the cell surface, an active process. For example, to stimulate the helper T (CD4) cells that promote the expansion and maturation of different immune effectors, antigen must be taken up, processed into peptides, and presented by specialized antigen-presenting cells (APCs) in the context of Class II major histocompatibility complex (MHC) antigens and second signals (Robinson and Delvig 2002). A protein that does not trigger uptake and presentation by APCs, as is the case for most self-proteins, may be nonimmunogenic yet still possess antigens. Under experimental conditions, these can be revealed by administering the protein together with an adjuvant to stimulate APC function (Schijns 2001). In reality, most natural immune responses are generated against invading pathogens and attributes of the infection likely provide the necessary immunogenic stimuli. Several toxins are among the limited group of noninfectious agents known to be highly immunogenic. Notably with respect to vaccine development, some of these remain immunogenic when their toxicity is attenuated.

Possible Advantages and Limitations of Using Plants to Produce Reagents for Active Vaccination

Plants as Expression Vectors

Edible plants expressing antigens that elicit protective immunity serve as the basis for the ideal vision of a vaccine that is inexpensive to both produce and deliver. While concerns about the impact of engineered plants on the environment and issues of profitability may dictate otherwise, the model plant-based vaccine could be grown locally using existing agricultural methods, harvested and fed to subjects. The plant approach would yield a reagent free of potential contaminants from the originating pathogen and other human or mammalian cell-based expression systems. Characteristics of the plant cells and tissues may be utilized to provide sufficient protection for the vaccine moiety to reach the small intestine for uptake and the induction of an immune response. Moreover, plant systems may be able to produce antigenic structures that are difficult to express in eukaryotic cell culture.

Plant expression systems have been successfully used to produce relatively complex proteins including structurally intact, active antibody molecules. These consist of two heavy and two light chains with a total molecular weight of approximately 150 kDa. Thus antigen complexity is not a concern and, at present, a low-yield and weak immunogenicity are the major limitations of plant antigen expression systems. There are a number of approaches being investigated to increase the yield of foreign proteins expressed in plants (reviewed in Gleba et al. 2005; Ko et al. 2003) and any obstacle here will likely be overcome in the near future. On the other hand, attempts to improve the immunogenicity of antigens expressed by plants have met with only partial success and this objective is not without concerns, as discussed in the following sections.

Plant Viruses and Bacteria as Expression Vectors

In a somewhat different but related approach to the use of engineered plants, vaccine antigens may be expressed by plant viruses or bacteria. In this case, plants are used to produce the agent, which can then be administered in purified form or using the infected plant tissue as a vehicle. The expectation here is that based on their structures, the plant virus or bacterium may share some immunogenic attributes but none of the pathological properties of human pathogens. While plant bacteria can express relatively complex antigens, the use of plant viruses as expression vectors may be restricted to relatively simple antigenic determinants because of issues with virus assembly. The inability to express intact antigenic structures has repercussions for the utility of the construct. For example, short peptide sequences expressed in an appropriate context may be able to induce a limited T cell response but are unlikely to trigger antibody production. Thus there would be no capacity to neutralize the target pathogen and the protective attributes, if any, would depend on a more rapid induction of a comprehensive immune response when an infection occurred. Yields of the vaccine antigen, in these expression systems only a fraction of the plant virus or bacteria, are also a concern. In addition, it is conceivable that the response to a weak antigen may be negatively impacted by prior tolerogenic exposure to native plant viruses and bacteria naturally present in the diet.

Our experience with the plant bacterium *Clavibacter xyli cynodontis* and an expression system using the coat protein of the alfalfa mosaic virus (AMV) expressed by tobacco mosaic virus highlights some of the differences between

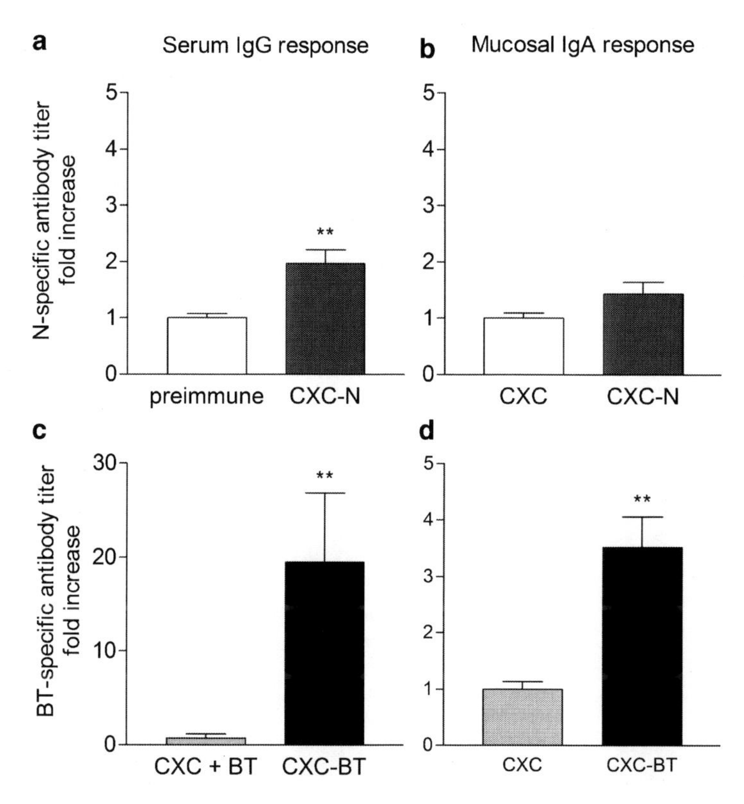

Fig. 1 Comparison of the immunogenicity of different *Clavibacter xyli cynodontis* constructs following parenteral and oral immunization. Groups of ten 8-week-old female Swiss-Webster mice were immunized with *Clavibacter xyli cynodontis* (*CXC*) engineered to express rabies virus nucleoprotein (N), *Bacillus thuringiensis* toxin (*BT*), or the native bacterium (*CXC*). Immunization consisted of a single dose of $5{\times}10^6$ CXC in saline intraperitoneally or two doses of $5{\times}10^7$ CXC in 10% sucrose at 2-week intervals *per os*. Serum and fecal pellet antibodies specific for N (**a, b**) and BT (**c, d**) were assessed 21 days after immunization by ELISA using antigen-coated plates and antibody isotype-specific secondary antibodies. The results are expressed as the fold-increase in antibody titers by comparison with preimmune levels. Statistically significant differences between the groups detected by the Mann-Whitney test are denoted by **, $p < 0.01$

these two approaches as well as shared limitations. We were able to express intact rabies virus nucleoprotein (54,000 MW) in *Clavibacter* and use this to readily elicit a nucleoprotein-specific immune response (Fig. 1). On the other hand, we were limited to expressing a peptide (N31D) derived from rabies nucleoprotein and glycoprotein in the AMV, achieving only relatively weak immunity (Yusibov et al. 2002). Nevertheless, both approaches were limited by the difficulty in obtaining high yields of material from infected plant tissues. For our *Clavibacter* experiments, the bacteria was grown *in vitro* and extensive purification of the virus from infected plant tissues was necessary to obtain a response in the AMV experiments. Any requirement for purification would place an additional financial constraint on the use of plant bacteria and viruses.

Basic Immunology: Routes of Vaccine Administration

Most existing vaccines are administered by injection into the dermis or muscle tissue. This primarily targets T-helper cells and a circulating IgG antibody response, even where a mucosal IgA antibody response may be more appropriate, but circumvents any difficulty with antigen stability in the gut. Vaccination targeting the naso-oropharynx may provide a means of inducing both IgA and IgG responses as well as reduced concerns about antigen stability but presents technical challenges with respect to administration. The oral route is clearly preferred. However, the antigens of an orally administered vaccine that targets the mucosa of the small intestine have to survive the environment of the gut and be taken up in an immunogenic form by the appropriate cells in the small intestine. The nature of the antigen and its carrier both contribute to this process. In our studies of antigens expressed by the plant bacterium *Clavibacter xyli cynodontis* (CXC), we compared the immune responses induced by intraperitoneal (i.p.) versus oral administration of two *Clavibacter* constructs: one expressing rabies virus nucleoprotein (N) and the second expressing *Bacillus thuringiensis* toxin (BT). Both were effective at inducing antibody responses to their respective antigens after i.p. injection but only a BT-specific response was seen after oral administration (Fig. 1). We expect that the inherently greater immunogenicity of the BT construct as well as the ability of BT antigens to survive the gut are important in this regard.

Plant expression systems may provide a natural means of preventing degradation of antigens by the gut. Antigen expression can be targeted to a variety of locations in different plant species and certain plant structures, plant cell walls, and plant cell organelles all have the potential to protect against antigenic degradation. For example, seeds may be particularly suited to the expression of high levels of antigen in a relatively stable package. While there has been extensive work on the expression of various vaccine antigens by plants, studies of the impact of targeting different plant structures on antigen stability are in early stages. Ideally, a plant-based structure that degrades and releases antigen in the small intestine can be identified and will function as a natural enteric capsule. However, a nonreplicating mucosal vaccine will primarily trigger a local secretory IgA response and T helper cell response. Cytotoxic T cells, an important arm of the immune response against pathogens invading across the mucosa, would not be efficiently induced. It remains to be proven for each pathogen whether or not such a limited response can limit the spread of infection and protect against disease.

Enhancing Immunogenicity of a Potential Plant Vaccine Versus the Risk of Breaking Tolerance to Food Antigens

Strategies to Make Antigens Expressed by Plants More Immunogenic

As a general rule, most noninfectious, nonreplicating antigens are not very immunogenic, the exceptions being structures that have unusual stimulatory properties such as toxins and superantigens. To induce an immune response under experimental conditions, an antigen is administered together with an adjuvant, a substance with properties that enhance the immune response to associated antigens. Adjuvants generally function by stimulating the activity of APCs such that the associated protein is efficiently taken up and presented in an immunogenic fashion to T cells. As our understanding of immunity has improved, a variety of structures that naturally have adjuvant-like properties have been identified including certain toxins (Lyche 2005; Choi et al. 2006), components of bacterial cell walls (Gustafson and Rhodes 1992; Jalava et al. 2003), DNA with CpG motifs (Krieg 2002), double-stranded RNA (Cui and Qiu 2006), and uric acid crystals released by dying cells (Shi et al. 2003). Strategies currently used to make antigens expressed by plants more immunogenic are, for the most part, based on those successful for other nonreplicating antigens, including the incorporation of an apathogenic toxin in the construct (Choe et al. 2006). However, there are other possibilities based on the differences in glycosylation between plants and other antigen expression systems. For example, proteins retained in the endoplasmic reticulum of plants undergo high mannose glycosylation (Ko et al. 2003) and mannose receptors are expressed by dendritic cells, the specialized APCs that drive primary immune responses (Diebold et al. 2002). Thus there is no doubt that the immunogenicity of antigens expressed in plants can be improved by several of the above approaches and an effective injectable vaccine can be produced *in planta*. However, as is the case for other noninfectious, nonreplicating oral vaccines, enhancing the immunogenicity of a plant-based oral vaccine is more problematic. The best characterized proteins with adjuvant activity following oral administration are the toxins of cholera and enteropathogenic *Escherichia coli* (Fujihashi et al. 2002). Considerable efforts have been made to dissociate the pathological attributes of cholera and LT toxins from their oral adjuvant properties, with varying success (Fujihashi et al. 2002). Subunits of these toxins can readily be expressed in the context of vaccine antigens in plant systems (Choe et al. 2006). Nevertheless, without other means of enhancing the immunogenicity of nonreplicating oral vaccines, repeat immunization using a toxin subunit may be limited, particularly if the toxin subunit is more immunogenic than the vaccine antigens.

Risks of Breaking Tolerance

Under normal circumstances, an ingested antigen that is taken up in the gut induces tolerance, an immunologically unresponsive state to the antigen. The existence of food allergies attests to the fact that certain food antigens can be taken up intact in the gut and induce an immune response in some individuals. One of the primary concerns of using plant-based systems for vaccination is that the vaccine construct may provoke a response to plant antigens, breaking tolerance and causing a food allergy. This is not simply a matter of a strong immune response occurring in the presence of food antigens or tolerance would be broken whenever someone gets a gastrointestinal infection. The prevalence of allergies to certain foods, for example peanuts, demonstrates that the nature of the antigen is an important risk element for the development of a food allergy. To induce a food allergy in experimental animals, a common food allergen must be administered with cholera toxin, a potent mucosal adjuvant (Helm et al. 2002). Since a nonreplicating oral vaccine likely must have adjuvant properties to be effective, there is a chance that sensitization against plant antigens contained in the vaccine may occur. The possibility that any such response would elicit the IgE antibodies commonly involved in an allergic reaction is more remote. In reality, there is probably as much of a chance that the mechanisms responsible for the maintenance of tolerance to plant-based food antigens may negatively impact the response to antigens expressed by the plant. Experiments with effective plant-based oral-mucosal vaccines are necessary to resolve these issues.

Other Uses of Plant-Based Immunological Reagents

Immunomodulation

While oral tolerance to experimental antigens has been studied for a number of years, relatively little is known about how tolerance to food antigens is generated and maintained. The possibility that the oral administration of antigen may induce tolerance or immune bias and reduce the severity of autoimmune disease has been investigated, with limited success, in multiple sclerosis (Faria and Weiner 2005).

However, the possibility that there may be a greater therapeutic effect if such antigens are administered in the context of plant antigens that are a normal part of the diet has not been thoroughly examined. Clearly, the recognition of antigen is required to induce antigen-specific antibodies and, as noted above, plant expression systems can be developed for enteric delivery of antigens. Further studies are needed to determine whether such systems may have added utility in the modification of aberrant immune responses in autoimmunity and food allergy.

Production of Antibodies for Passive Immunization in Plants

Passive immunization, the rapid provision of protection to an individual by the administration of preformed, antigen-specific antibodies, is extensively used in a variety of situations ranging from the postexposure treatment of infections such as rabies (Hanlon et al. 2001) to the prevention of Rh sensitization after an Rh-negative mother has delivered an Rh-positive child (Kumpel 2002). Passive immunization may prove to be the most effective means of protecting a population after a bioterror attack in that the delay required for the production of protective levels of antibody after active immunization is eliminated and the level of protection provided when necessary is uniform. Extensive stocks of antibodies are required for these applications as well as the use of passive immunization in immunocompromised individuals. At present, most antibodies used for passive immunization are isolated from the sera of immunized individuals, limiting their supply, and presenting a potential risk of contamination with human pathogens. Monoclonal antibodies are now being investigated as replacements for the more crude immunoglobulin preparations for certain applications (Hanlon et al. 2001; Sawyer 2000; Keller and Stiehm 2000). However, the production of monoclonal antibodies using conventional culture technologies is expensive and often does not eliminate the possibility of contamination with a pathogen. As noted above, intact, functional antibody molecules have been successfully expressed in plants (Ko et al. 2003; Verch et al. 1998) and plant-based technologies may prove advantageous over conventional antibody production methods in cost and safety.

Conclusions

There is no doubt that plants can be used as factories to produce immunological reagents and that such products will become widely available in the foreseeable future. Any nonreplicating injectable vaccine could be expressed in planta. However, as is the case for any noninfectious oral vaccine, the prospects of developing plant-based edible vaccines are more difficult to predict. To be successful, several elements of the system must come together: (1) the nature and level of expression of the antigen; (2) the effectiveness of the plant tissues as a delivery vehicle to protect against antigen degradation in the gut; (3) the capacity of the construct to promote antigen uptake in the gut; and (4) the immunogenicity of the construct. Each new construct studied that addresses any of these issues provides us with a better understanding of the complex interactions between the immune system and ingested antigens and closer to the ultimate goal of a safe, inexpensive, and effective plant-based oral vaccine.

Acknowledgements The author thanks Laura Conway for the transgenic *Clavibacter*, Anna Modelska for conducting the experiments, as well as Christine Brimer and Rhonda Kean for assistance in preparing the manuscript. This work was supported by a grant to the Biotechnology Foundation Laboratories Inc. from the Commonwealth of Pennsylvania.

References

Bodey B, Bodey B Jr, Siegel SE, Kaiser HE (2000) Genetically engineered monoclonal antibodies for direct anti-neoplastic treatment and cancer cell specific delivery of chemotherapeutic agents. Curr Pharm Des 6:261–276

CDC (2002) General recommendations on immunization. Recommendations of the Advisory Committee on Immunization Practices (ACIP) and the American Academy of Family Physicians (AAFP). MMWR Mort Morbid Wkly Rep 51(RR02):1–36

Choe NW, Estes MK, Langridge WH (2006) Synthesis of a ricin toxin B subunit-rotavirus VP7 fusion protein in potato. Mol Biotechnol 32:117–128

Choi NW, Esters MK, Langridge WH (2006) Ricin Toxin B subunit enhancement of rotavirus NSP4 immunogenicity in mice. Viral Immunol 19:54–63

Cohen-Kaminsky S, Jambou F (2005) Prospects for a T-cell receptor vaccination against myasthenia gravis. Expert Rev Vaccines 4:473–492

Cui Z, Qiu F (2006) Synthetic double-stranded RNA poly (I:C) as a potent peptide vaccine adjuvant: therapeutic activity against human cervical cancer in a rodent model. Cancer Immunol Immunother 55:1267–1279

Diebold SS, Plank C, Cotton M, Wagner E, Zenke M (2002) Mannose receptor-mediated gene delivery into antigen presenting dendritic cells. Somat Cell Mol Genet 27:65–74

Dredge K, Marriott JB, Todryk SM, Muller GW, Chen R, Stirling DI, Dalgleish AG (2002) Protective antitumor immunity induced by a costimulatory thalidomide analog in conjunction with whole tumor cell vaccination is mediated by increased Th1-type immunity. J Immunol 168:4914–4919

Faria AM, Weiner HL (2005) Oral tolerance. Immunol Rev 206:232–259

Fujihashi K, Koga T, van Ginkel FW, Hagiware Y, McGhee JR (2002) A dilemma for mucosal vaccination: efficacy versus toxicity using enterotoxin-based adjuvants. Vaccine 20:2431–2438

Gleba Y, Klimyuk V, Marillonnet S (2005) Magnification-a new platform for expressing recombinant vaccines in plants. Vaccine 23:2042–2048

Gustafson GL, Rhodes MJ (1992) Bacterial cell wall products as adjuvants: early interferon gamma as a marker for adjuvants that enhance protective immunity. Res Immunol 143:483–488

Hanlon CA, DeMattos CA, DeMattos CC, Niezgoda M, Hooper DC, Koprowski H, Notkins A, Rupprecht CE (2001) Experimental utility of rabies virus-neutralizing human monoclonal antibodies in post-exposure prophylaxis. Vaccine 19:3834–3842

Helm RM, Furuta GT, Stanley JS, Ye J, Cockrell G, Connaughton C, Simpson P, Bannon GA, Burks AW (2002) A neonatal swine model for peanut allergy. J Allergy Clin Immunol 109:136–142

Jalava K, Eko FO, Riedmann E, Lubitz W (2003) Bacterial ghosts as carrier and targeting systems for mucosal antigen delivery. Expert Rev Vaccines 2:45–51
Jenner E (1798) An inquiry into the causes and effects of the variolae vaccine. London: printed for the author by Sampson Law
Keller MA, Stiehm ER (2000) Passive immunity in prevention and treatment of infectious diseases. Clin Microbiol Rev 13:602–614
Ko K, Tekoah Y, Rudd PM, Harvey DJ, Dwek RA, Spitsin S, Hanlon CA, Rupprecht C, Dietzschold B, Golovkin M, Koprowski H (2003) Function and glycosylation of plant-derived antiviral monoclonal antibody. Proc Natl Acad Sci U S A 100:8013–8018
Ko K, Steplewski Z, Glogowska M, Koprowski H (2005) Inhibition of tumor growth by plant-derived mAb. Proc Natl Acad Sci U S A 102:7026–7030
Krieg AM (2002) CpG motifs in bacterial DNA and their immune effects. Annu Rev Immunol 20:709–760
Kumpel BM (2002) On the mechanism of tolerance to the Rh D antigen mediated by passive anti-D (Rh D prophylaxis). Immunol Lett 82:67–73
Lenarczyk A, Le TT, Drane D, Malliaros J, Pearse M, Hamilton R, Cox J, Luft T, Gardner J, Suhrbier A (2004) ISCOM based vaccines for cancer immunotherapy. Vaccine 22:963–974
Li ZG, Mu R, Dai ZP, Gao XM (2005) T cell vaccination in systemic lupus erythematosus with autologous activated T cells. Lupus 14:884–889
Lyche N (2005) Targeted vaccine adjuvants based on modified cholera toxin. Curr Mol Med 5:591–597
Pasteur L (2002) Summary report of the experiments conducted at Pouilly-le-Fort Near Melun, on the anthrax vaccination. Classics of Biology and Medicine. Yale J Biol Med 75:59–62
Ramon G, Zoeller C (1927) L'anatoxine tétanique et l'immunisation active de l'homme vis-à-vis du tétanos. Ann Inst Pasteur 41:803–833
Robinson JH, Delvig AA (2002) Diversity in MHC class II antigen presentation. Immunology 105:252–262
Sawyer LA (2000) Antibodies for the prevention and treatment of viral diseases. Antiviral Res 47:57–77
Schijns VE (2001) Induction and direction of immune responses by vaccine adjuvants. Crit Rev Immunol 21:75–85
Shi Y, Evans JE, Rock KL (2003) Molecular identification of a danger signal that alerts the immune system to dying cells. Nature 425(6957):516–521
Vandenbark AA, Morgan E, Bartholomew R, Bourdette D, Whitham R, Carlo D, Gold D, Hashim G, Offner H (2001) TCR peptide therapy in human autoimmune diseases. Neurochem Res 26:713–730
Verch T, Yusibov V, Koprowski H (1998) Expression and assembly of a full-length monoclonal antibody in plants using a plant virus vector. J Immunol Methods 220:69–75
Vichier-Guerre S, Lo-Man R, BenMohamed L, Deriaud E, Kovats S, Leclerc C, Bay S (2003) Induction of carbohydrate-specific antibodies in HLA-DR transgenic mice by a synthetic glycopeptide: a potential anticancer vaccine for human use. J Pept Res 62:117–124
Weinstein PD, Cebra JJ (1991) The preference for switching to IgA expression by Peyer's patch germinal center B cells is likely due to the intrinsic influence of their microenvironment. J Immunol 147:4126–4135
Yusibov V, Hooper DC, Spitsin SV, Fleysh N, Kean RB, Mikheeva T, Deka D, Karasev A, Cox S, Randall J, Koprowski H (2002) Expression in plants and immunogenicity of plant virus-based experimental rabies vaccine. Vaccine 20:3155–3164

Display of Peptides on the Surface of Tobacco Mosaic Virus Particles

M.L. Smith, W.P. Fitzmaurice, T.H. Turpen, and K.E. Palmer

Contents

Abstract In this review, we focus on the potential that tobacco mosaic virus (TMV) has as a carrier for immunogenic epitopes, and the factors that must be considered in order to bring products based on this platform to the market. Large Scale Biology Corporation developed facile and scaleable methods for manufacture of candidate peptide display vaccines based on TMV. We describe how rational design of peptide vaccines can improve the manufacturability of particular TMV products. We also discuss downstream processing and purification of the vaccine products, with particular attention to the metrics that a product must attain in order to meet criteria for regulatory approval as injectable biologics.

Keywords tobacco mosaic virus, vaccine, plant, virus-like particle, papillomavirus

---

M.L. Smith
Genentech, Inc. 1000 New Horizons Way Vacaville, CA 95688, USA

W.P. Fitzmaurice
Novici Biotech., Vacaville, CA 95688, USA

T.H. Turpen
Technology Innovation Group, Inc., 7006 Firewheel Hollow, Austin, TX 78750, USA

K.E. Palmer(✉)
Owensboro Cancer Research Program, University of Louisville, 1020 Breckenridge Street, Owensboro, KY 42303, USA
e-mail: kenneth.palmer@louisville.edu

A.V. Karasev (ed.) *Plant-produced Microbial Vaccines.*
Current Topics in Microbiology and Immunology 332

Introduction

It is now well established that immunogenic peptides are most efficiently presented to the immune system in a highly ordered, repetitive, quasicrystalline array such as by a virus-like particle (VLP). By their structure, some VLPs are thought to be capable of stimulating proliferation of dendritic cells and other antigen-presenting cells. Thus, antigen-specific B and T cell responses are markedly enhanced when epitopes are coupled to VLPs. The regular array of epitopes on the surface of chimeric VLPs is thought to allow efficient crosslinking of antigen-specific immunoglobulins on B cells, leading to B cell proliferation and production of antibodies. Exposure of the immune system to a repetitive array of self-peptides is an effective mechanism of breaking immunological tolerance and producing auto-reactive therapeutic antibodies. Therefore, there is currently significant interest in VLP-epitope display systems for induction of antibodies for disease therapy and prophylaxis, as well as for induction of peptide-specific T cell responses for immunotherapy of cancer and chronic disease. There are well-established methodologies for recombinant production of VLP-epitope display systems that use self-assembling capsid proteins of several different viruses, most notably papillomaviruses, hepatitis B core and surface antigens, and various bacteriophages, including the leviviruses MS2 and Qβ. Many of these are easily produced in various eukaryotic expression systems, and some may also be manufactured in bacteria. The scientific literature is replete with examples of plant-produced recombinant virus particle and VLP epitope display-based vaccines that show utility, mainly as prophylactic vaccines against infectious diseases.

Research on the use of recombinant virus-like particles as epitope carriers started in the mid-1980s, where in addition to tobacco mosaic virus (TMV) (Haynes et al. 1986), hepatitis B surface and core antigen particles, yeast Ty Gag particles and poliovirus virions were shown to be effective carriers that significantly enhanced the immunogenicity of linked epitopes (Valenzuela et al. 1985; Delpeyroux et al. 1986; Adams et al. 1987; Clarke et al. 1987; Burke et al. 1988; Delpeyroux et al. 1988; Martin et al. 1988, 2003; Clarke et al. 1990). However, VLP epitope display technology has only recently moved toward practical development of vaccines for use in human and veterinary medicine. It is interesting to note that two promising malaria vaccines based on epitope display on the surface of VLPs are being tested in humans. The product that is most advanced in clinical evaluation is a promising malaria vaccine based on display of a large part of the malaria parasite *Plasmodium falciparum* circumsporozoite protein, in chimeric recombinant HBsAg particles. This vaccine, RTS,S, which is being developed by GlaxoSmithKline Biologicals, shows promising efficacy in human vaccinee volunteers challenged with *P. falciparum*, and a phase III efficacy trial in children in Mozambique is planned for the near future (Moorthy et al. 2004). The Malaria Vaccine Initiative, in collaboration with biotechnology company Apovia Inc., is conducting a phase I trial with recombinant HBcAg particles displaying a B cell epitope from the circumsporozoite protein fused to the major immunogenic domain of the core protein, and a universal T-helper epitope at the carboxy-terminus (Birkett et al. 2002; Moorthy et al. 2004). These are good examples of how

VLP-epitope display products are starting to show their promise for addressing infectious diseases. Thus far, only one plant VLP-based candidate vaccine has been tested in humans: a candidate rabies virus vaccine (Yusibov et al. 2002), and as far as we are aware there are no plant-produced VLP products that are nearing commercialization. The focus of this review is on some of the practical issues that must be considered in order to bring plant-produced virus particle and VLP-based epitope display systems into commercial use. In light of the fact that our primary interest is in TMV-based VLP and virion peptide display, we will concentrate our discussions on the TMV platform technology, and review some of the scientific, manufacturing, and regulatory issues that must be addressed in order to commercialize this technology.

Overview of Technologies for Display of Peptides on the Surface of Tobacco Mosaic Virus

TMV has been studied as a model antigen for over 50 years (reviewed in Van Regenmortel 1999). The viral particles are excellent immunogens, and much of the foundation of modern immunology was laid by TMV serological research, for example, neutralization of virus infectivity by antibodies was demonstrated in the TMV system 15 years before similar results were obtained in animal viral systems. Years before the mechanism of presentation of antigen to cells of the adaptive immune system by professional antigen-presenting cells (APCs) was understood, Loor (1967) demonstrated in rabbits that 14C-labeled TMV virions were rapidly and effectively transported from the site of injection to proximal lymph nodes and then to the spleen. With remarkable foresight, both Loor (1967) and Marbrook and Matthews (1966) showed that the particulate nature of viruses is essential for their immunogenicity because disassembled plant virus coat protein induced lower titer antibodies and poorer immune memory in animals immunized with equivalent doses of intact and disassembled viruses. In the ^{14}C-labeling experiments referred to above, Loor also showed that the particulate nature of viruses is important for uptake by and activation of APCs since far lower amounts of labeled coat protein were transported by APCs to lymphoid organs in animals immunized with disassembled virus in comparison with intact virus particles (Loor 1967).

The first demonstration that VLPs derived from TMV coat protein could display antigenic epitopes was reported by Haynes et al. (1986), who displayed a poliovirus epitope on the surface of VLPs assembled from TMV coat protein expressed in *Escherichia coli*. With the advent of the first infectious clones of TMV (Dawson et al. 1986; Meshi et al. 1986), it became possible to manipulate the genome in vitro and thus to construct recombinant TMV vectors and fuse epitopes to the surface of the coat protein. These technologies were the foundation of a biotechnology company, now called Large Scale Biology Corporation (LSBC), dedicated to production of vaccines and therapeutics for human and animal health through exploitation of recombinant TMV (reviewed by Turpen 1999). Since 1987, we have developed and successfully implemented technologies and infrastructure for rapid, cost-effective expression and extraction of recombinant proteins from plants, including the world's only operating, commercial-scale biomanufacturing facility for recombinant protein extraction from plant tissues. Commercial-scale biomanufacturing allows translation of the body of data providing proof of the concept that TMV can function as an effective carrier for antigenic peptides into vaccine products for human and veterinary applications (Pogue et al. 2002).

The tobamovirus virion is a rigid rod of about 18 nm in diameter and 300 nm in length. The structures of the virion and coat protein have been determined by x-ray diffraction (Watson 1954; Namba and Stubbs 1986). The virion contains approximately 2,130 coat protein subunits, each approximately 17.5 kDa, arranged in a right-handed helix with 16.3 subunits per turn. Each subunit of the coat protein tolerates insertion of epitopes at one of three solvent-exposed positions: the N-terminus, at or near the C-terminus and in a surface-exposed loop corresponding to amino acids 59–65. The density with which epitopes may be displayed on rod-shaped viruses such as TMV is unmatched by any competing VLP system. For example, it is possible to display greater than 2,100 copies of an epitope on the surface of TMV, compared to 180 on a T=3 particle such as HBcAg VLP or 420 on a T=7 particle such as papillomavirus VLP. Table 1 summarizes the most significant reports of display of immunogenic peptides on the surface of TMV.

Table 1 Vaccine epitopes produced on the surface of TMV

Vaccine model	Summary of results	Reference
Malaria B cell epitope	Successful display of epitope from malaria on the surface of TMV. High yield in field production	Turpen et al. 1995
Mouse zona pellucida ZP3 epitope, a model contraceptive vaccine	Vaccination was able to break B cell tolerance and induce autoreactive antibodies that recognized zona pellucida located in mouse ovaries. This study is the first and only published account of the use of a plant virus particle to break immune tolerance	Fitchen et al. 1995
Murine hepatitis coronavirus neutralizing epitope	Five of six vaccinated mice were protected from challenge with murine hepatitis virus	Koo et al. 1999
Feline panleukopenia parvovirus epitope	Vaccinated cats were partially protected against challenge with feline panleukopenia parvovirus	Pogue et al. 2004
Pseudomonas aeruginosa OMPF peptide	Vaccinated mice had reduced lesion number and disease severity when challenged with *Pseudomonas*	Staczek et al. 2000
Foot and mouth disease virus neutralizing epitope	Guinea pigs and swine protected against challenge with FMDV	Wu et al. 2003
Rabbit papillomavirus L2 Imer epitopes	Full protection against homologous virus challenge; some cross-protective immunity between papillomavirus types	Palmer et al. 2006
Mouse tumor T Cell epitopes	Induced T cell responses that protected mice against terms challenge	McCormick et al. 2006

Overall, these data show that TMV-based peptide display vaccines can be applied for prevention of infectious disease. Of all of these publications, the results presented by Fitchen et al. (1995) are perhaps the most important, since they show that TMV particles displaying a repetitive array of self-peptides can break B cell tolerance and allow production of autoreactive antibodies. In the following sections, we propose some solutions to the manufacturing and regulatory issues associated with bringing TMV, and TMV-based VLP carriers into commercial production.

Manufacture and Purification of Tobacco Mosaic Virus Particle-Based Vaccines

To date no vaccine based on a chimeric plant virus has been delivered by parenteral injection to human subjects. However, LSBC has developed a patient-specific vaccine for the treatment of non-Hodgkin's lymphoma (NHL) (McCormick et al. 2003). The idiotype regions of the tumor-specific IgG were successfully expressed as single chain Fv proteins in tobacco plants using the TMV-based GENEWARE expression platform. These vaccines were tested in a phase I clinical trial with 16 patients, and an excellent safety profile was demonstrated, together with encouraging immune response profiles. This study was the first to test a plant-derived biologic delivered parenterally (McCormick et al. 2008). The quality control (QC) and quality assurance (QA) framework implemented for this phase I trial will facilitate the transition of plant virus vaccines from the laboratory to evaluation in humans.

The patient-specific nature of the NHL vaccine lent itself to a manufacturing process on the milligram scale, with purification from growth room-cultivated plants. In contrast, for vaccines targeting human and animal pathogens, a yield of 1–10 kg of final product per manufacturing run can be anticipated. In a recent review (Pogue et al. 2002), production at this scale was discussed, in addition to the regulatory issues surrounding the use recombinant TMV vectors in field-based production. Herein we will focus on the processing and purification considerations and challenges that must be addressed, to permit TMV fusions to be produced and formulated as human and veterinary vaccines. We will use data from a number of LSBC investigational vaccine programs to illustrate key points.

We carry out the majority of the developmental work associated with a new coat protein fusion with *Nicotiana benthamiana* as the host plant for the recombinant virus, since *N. benthamiana* is very susceptible to TMV infection and is easily cultivated under growth-room conditions. At scale and under field conditions, *N. benthamiana* is a suboptimal host as its growing season is limited to the cooler months because elevated summer temperatures adversely affect biomass yield (Fig. 1).

Some cultivated varieties of tobacco are highly susceptible to wild-type TMV, but the yield of recombinant virion can be reduced significantly when foreign epitopes are displayed on the coat protein surface. The requirement for host susceptibility coupled with acceptable biomass yield under field conditions prompted LSBC to embark on a breeding program to develop improved plant hosts (Fitzmaurice 2002).

Fig. 1 a–c Plant host growth characteristics under field conditions. **a** Comparison of biomass yield for an improved host, *N. excelsiana* (*#1*) and two different accessions of *N. benthamiana* (*#2*, plants mostly dead and *no. 3*, grown under field conditions in Kentucky, mid-summer). **b** Close-up of *N. excelsiana* (*#1*), 13 days after transplanting, cultivated during the spring. **c** Close-up of *N. benthamiana* (*#3*), 28 days after transplanting, cultivated during the spring. Plants shown in **b** and **c** were cultivated in Kentucky

Fig. 2 a–c Effect of cultivation conditions on *N. excelsiana* plant morphology. **a** Field-grown plants, harvested 52 days after transplanting. **b** Growth room cultivated plant, 48 days after sowing. **c** Greenhouse grown plant, 43 days after sowing

The improved performance of the alternative plant hosts upon field cultivation is clear. Figure 2 illustrates the superior growth characteristics of a novel LSBC-proprietary host (*N. excelsiana*), an interspecific hybrid between *N. excelsior* and *N. benthamiana*, in the field, greenhouse and under growth-room conditions, at a typical age of harvest.

Since growing conditions clearly alter the host plant morphology, the effect of the growth environment on host susceptibility to TMV, as well as product yield and quality, must be addressed. In particular, the host response to the pathogen will differ with the growth conditions. One outcome of this may be altered accumulation and stability of epitope fusion virions due to the complement of proteases upregulated within the host upon infection. For example, screening for fusion accumulation in different hosts, cultivated under growth-room conditions, illustrated that epitope stability was a function of the particular host/epitope combination (Fig. 3a). Field growth conditions may also impact this relationship. Promising vaccine candidates whose production requirements dictate production in the field should, therefore, be evaluated under actual growth conditions early in the product development cycle. The availability of a number of virus-susceptible field-adapted cultivars for screening is clearly desirable, should the initially selected host prove suboptimal. LSBC has a portfolio of production host candidates that may be screened for optimal performance with particular coat protein fusion vectors (Fitzmaurice 2002). By being cognizant of the biological parameters that affect the relationship between a recom-

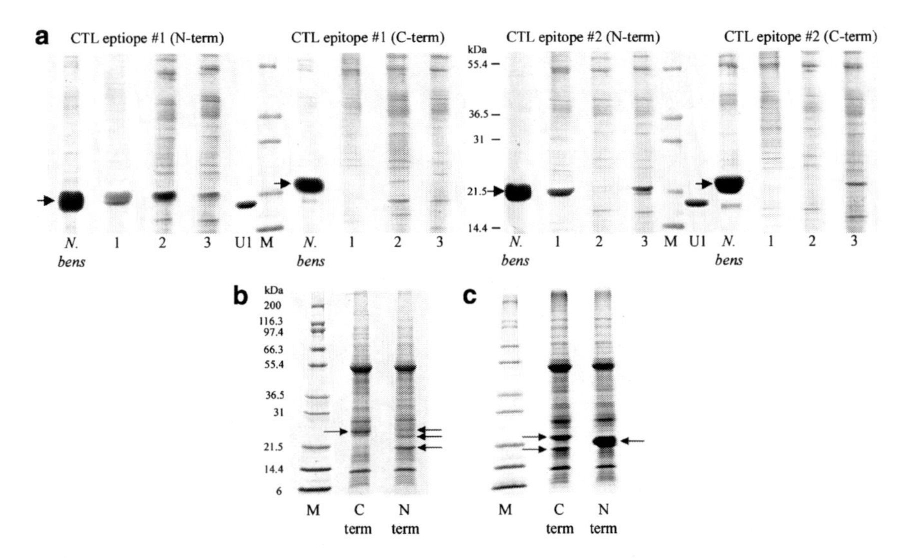

Fig. 3 a–c TMV coat protein fusion stability and expression as a function of the plant host employed and fusion location. **a** Screening of partially purified TMV peptide fusions from *N. benthamiana (N. bens)* and a series of field-adapted cultivars (*1, 2* and *3*). Plant host was screened under growth room conditions. Two 9 amino acid HLA-A2*01-restricted cytotoxic T lymphocyte (*CTL*) epitopes (CTL epitopes *#1* and *#2*) were tested as fusions to either the N-terminus (*N-term*) or located at the C-terminus of coat protein, internal to the last four residues (*C-term*). The coat protein fusion for each is identified by an *arrow*. *M* Mark 12 molecular-weight ladder, *U1* wild-type U1 TMV control. **b** Example of peptide fusion which showed stability when located at the C-terminus of coat protein, internal to the last four residues, but was subject to proteolytic cleavage in vivo at the N-terminus. **c** Example of in vivo peptide fusion cleavage when located at the C-terminus of coat protein, internal to the last four residues, while peptide stability was obtained as an N-terminal fusion. For **b** and **c** the identity of both the full-length and the truncation products (identified by *arrows*) was confirmed by MALDI-TOF (matrix-assisted laser desorption/ionization-time of flight) analysis

binant virus and the host, we have been able to manipulate the system to ensure that any given product can be manufactured in an optimal fashion.

The stability of the peptide displayed on TMV is also dependent on the location of the fusion peptide in the coat protein. Published reports have focused predominantly on peptide display at the C-terminus and within the surface loop, with certain studies favoring peptide insertion preceding the four C-terminal residues, presumably to prevent proteolytic cleavage (Fitchen et al. 1995; Sugiyama et al. 1995; Turpen et al. 1995; Koo et al. 1999; Staczek et al. 2000; Wu et al. 2003). During preliminary screening for expression, we recommend that fusion to the N-terminus of TMV U1 also be considered. Although cases have been noted where the internal C-terminus location prevented in vivo proteolytic degradation when compared to N-terminal display, we have also observed the reverse (Fig. 3b, c).

Yet a further consideration is the influence of the peptide fusion on the virus–host interactions. For TMV, epitopes with a high isoelectric point and positive charge were found to result in a necrotic response by the host (Bendahmane et al. 1999). A high isoelectric point was also shown to be deleterious for the systemic infection of cowpea mosaic virus fusions (CPMV), with epitope length being another factor (Porta et al. 2003). Both studies noted that the addition of acidic residues counterbalanced the undesirable properties associated with the epitope fusions. The beneficial effect of compensatory acidic residues has also been noted for TMV-displayed epitopes in our laboratory (Fig. 4). The addition of compensatory residues can, therefore, be viewed as a strategy to rescue the expression of certain coat protein

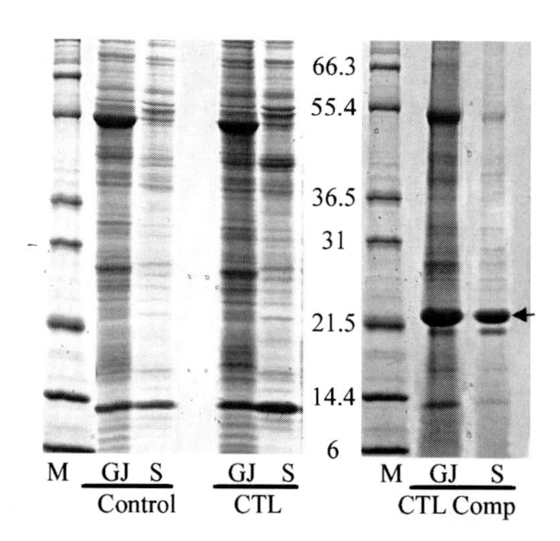

Fig. 4 Recovery of TMV coat protein fusion expression by the addition of compensatory acidic acid residues. A mouse CTL epitope was expressed as an N-terminal fusion to the U1 coat protein. The epitope contained one basic residue and no acidic residues. *M* Mark 12 (Invitrogen) molecular-weight marker, *GJ* starting green juice (crude extract), *S* Supernatant following green juice processing, *Control* Uninfected *N. benthamiana* control, *CTL* original CTL epitope coat protein fusion, *CTL Comp* CTL epitope coat protein fusion with compensatory residues. *Arrow* indicates coat protein fusion, the identity of which was confirmed by MALDI-TOF analysis. For the GJ and S CTL Comp, samples were diluted 5× prior to loading. For S Control and S CTL, samples were loaded at 1×

fusions. However, examples exist for the successful expression of positively charged peptides on the surface of TMV (Wu et al. 2003), suggesting that the peptide–host relationship is more nuanced than initially proposed (Bendahmane et al. 1999).

Multiple procedures have been developed for the purification of viruses from infected plant extracts on a laboratory scale (Corbett 1961; Dunn and Hitchborn 1965; Timian and Savage 1966; Gooding and Hebert 1967). Several of these employ an organic extraction or high-speed centrifugation step, which are undesirable when processing is done on a production scale (2,000–4,000 kg infected tissue per day). Large Scale Biology Corporation has developed a method compatible with the processing of large masses of plant material (Garger et al. 2000, 2001). The process was validated through the purification of a TMV coat protein fusion displaying a malarial epitope (Turpen et al. 1995; Pogue et al. 2002). This procedure is illustrated schematically in Fig. 5a. Briefly, the infected plant tissue was harvested and homogenized in the presence of 0.5 l water/kg of biomass. Removal of the plant fiber yielded a crude extract, or "green juice" that was adjusted to pH 5 and heated to 45–50°C. After incubation for approximately 10 min, the juice was cooled to below 10°C and a low-speed centrifugation step employed to separate the coagulated Fraction I proteins from the virus-containing supernatant. Ultrafiltration was employed to concentrate the supernatant 40- to 50-fold, prior to virus precipitation by the addition of polyethylene glycol (*M*r 8,000) and sodium chloride, each at 4% w/v. The precipitated virus was recovered as a paste by centrifugation, with the soluble host proteins remaining in the discarded supernatant. Operating at scale, this process yielded 0.6–1 kg of purified TMV fusion product per acre of infected plant material. The final purity of this TMV malarial epitope fusion was greater than 95% (Pogue et al. 2002). To evaluate the generalizability of this process, we have used a scaled-down version of the procedure, which faithfully mirrors the principal parameters for the production process, e.g., centrifugation conditions and holding times at the different temperatures. The process (outlined in Fig. 5a) has yielded good recoveries, with final purities comparable to the malarial epitope coat protein fusion, for 30%–40% of the peptide fusions evaluated. The remaining epitope fusions partitioned with the Fraction I proteins following centrifugation of the pH-adjusted and heat-treated green juice (Fig. 5b). These fusions can be recovered by resuspension of the Fraction I pellet under slightly alkaline conditions, followed by an additional centrifugation step. These process modifications, which are readily scaleable (Garger et al. 2000, 2001), effectively release the TMV fusion into the supernatant, with good recoveries, while the Fraction I proteins remain in an aggregated form. The processing of this supernatant can then proceed as indicated in Fig. 5A and yields a final product that meets purity requirements.

The displayed peptide may alter the purification characteristics of the recombinant virus at other stages of the process as well. For example, we have noted cases where the peptide prevents complete precipitation of the virus following the addition of NaCl and polyethylene glycol (Fig. 5c), a condition that can be remedied by the alteration of the ionic strength and polyethylene glycol concentration employed. The resistance of each fused peptide to proteases is an additional characteristic that must be evaluated. In vivo proteolysis has been alluded to above (Fig. 3); however, this is

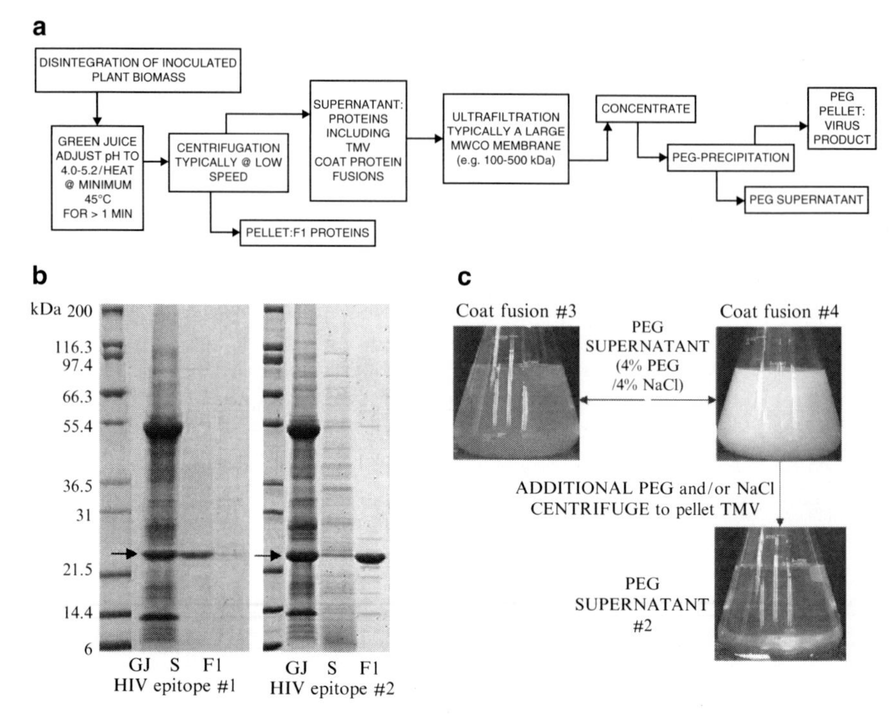

Fig. 5 **a** Simplified process flow diagram for the extraction and purification of TMV coat protein fusions on a multimetric ton per day scale (adapted from Garger et al. 2000). **b** Illustration of the differential partitioning observed between the pH adjusted and heat-treated supernatant and the coagulated Fraction I proteins for two TMV coat protein fusions, displaying peptides derived from HIV envelope proteins. *GJ* initial green juice (crude extract), *S* supernatant following green juice processing, *F1* processed F1 fraction. *Arrow* indicates TMV coat protein fusion. **c** Incomplete precipitation of TMV fusions under standard NaCl/polyethylene glycol (4% w/v each) conditions. Image shows the PEG supernatants (see **a**) following the centrifugation step to precipitate the virus fusion. Coat fusion *#3* precipitated as expected, leaving a clear supernatant, while incomplete precipitation was observed with coat fusion *#4*. For coat fusion *#4*, higher polyethylene glycol concentrations were required to obtain full precipitation of the virus, leaving a clear supernatant (PEG SUPERNATANT *#2*)

not necessarily predictive of stability during processing, as the loss of cellular compartmentalization with homogenization will expose the TMV fusion to additional proteolytic activities. One approach consists of incubating in-process samples, e.g., the initial supernatant, at 20–22°C (room temperature) for 2 h and evaluating stability by gel analysis, upon completion of the purification. For the peptide fusion shown in Fig. 6, excellent stability was obtained at bench-scale under standard processing conditions, where temperatures were maintained below 10°C (with the exception of the heat-treatment step). In contrast, with the room temperature incubation, substantial truncation was observed. By defining the process parameters critical to the stability of a particular fusion, the appropriate precautions can be implemented during scaled-up production, where the processing and holding times encountered

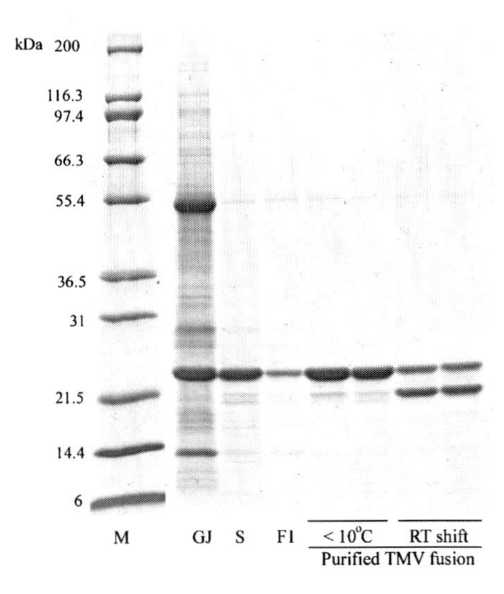

Fig. 6 Evaluation of TMV peptide fusion stability during processing. *M* Mark 12 (Invitrogen) molecular weight marker, *GJ* starting green juice, *S* supernatant following green juice processing, *F1* processed F1 fraction; *<10°C* purified TMV fusion (displaying a peptide from the L2 protein of human papillomavirus) obtained from S, with processing steps performed under chilled conditions, *RT shift* purified TMV fusion obtained from the supernatant S, which was incubated for 2 h at room temperature prior to processing under standard conditions

are typically of a longer duration. In summary, while each TMV peptide fusion modifies the particle surface properties and consequently its processing and purification characteristics, our experience has been that with the appropriate modifications, the majority of fusions evaluated can be made to conform to a general and scaleable production process, yielding a final product of acceptable purity.

For those cases where limited solubility persists despite processing modifications, alterations to the peptide itself may prove beneficial. Koo et al. (1999) noted that TMV displaying a ten amino acid peptide from the S2 glycoprotein of murine hepatitis virus was relatively insoluble, even though the recombinant coat protein expressed effectively and formed virus particles. When this peptide was flanked by additional S2 glycoprotein sequence, the TMV fusion was soluble and readily purified. This suggests that tiling around problematic peptides of interest may yield sequences with improved solubility characteristics when displayed on the surface of TMV. An alternative approach would be the introduction of compensatory sequences, in a manner similar to that described earlier in the context of improving virus–host compatibility, see for example Smith et al. (2006). For both approaches, immunogenicity testing in the appropriate model systems would be required to ensure the altered TMV fusions retain their desired properties.

Final Product Release and Stability

Once the bulk purified vaccine has been obtained, it must be formulated with excipients and/or adjuvants, sterilized and filled (Patro et al. 2002). The final form of the vaccine must also be considered. Is it to be distributed as a liquid or as a lyophilized powder, to be reconstituted with diluent prior to use? The filled product must also be submitted to extensive stability testing under the intended storage conditions to ensure that the product retains both integrity and immunogenicity.

There are a great number of potential excipients compatible with the formulation of parenteral drugs (Powell et al. 1998). For TMV fusions, which can be resuspended following purification in the required formulation buffer, e.g., phosphate buffered saline or Hank's buffered saline solution, formulation normally involves only a protein concentration adjustment and may include the incorporation of an adjuvant. However, issues relating to sterilization and TMV inactivation, discussed below, will impact the formulation steps, in particular relating to final product pH, ionic strength, and possibly the timing of adjuvant incorporation. An area where additional excipient testing may be of interest relates to the solubility of the TMV fusion. Wild-type TMV is inherently soluble and can be stored at concentrations above 20 mg/ml for several years at 4°C, with no visible signs of aggregation, degradation, or precipitation. However, the solubility characteristics of the virus can be changed dramatically by the addition of peptide fusions, and we have noted several examples where visible precipitation occurs with storage at 4°C, even at relatively low concentration (1–2 mg/ml). Addition of the appropriate excipient may prevent this from occurring, and we have found that aggregation state can be modulated and controlled through addition of excipients. Protein–protein aggregation is typically viewed as a negative for protein biologics since it may adversely affect activity and half-life (Patro et al. 2002). The impact, if any, of aggregation on a vaccine product needs to be considered. Given that aggregates can enhance the immune response, the outcome may ultimately be a beneficial one if potency is controlled.

Following the formulation of protein biologics, sterile filtration through a 0.2 μm filter is typically performed, to ensure the product is free from bioburden. However, the physical dimensions of TMV (18×300 nm) results in the rapid fouling of 0.2-μm filters. Although the methods that we have developed for purification of TMV generally yield product with very low endotoxin and bioburden loads, the implementation of a method or combination of methods to permit TMV fusion vaccine sterilization, to eliminate residual bioburden, is a priority. One possibility is serial 0.45-μm sterile filtration, since 0.2-μm filtration is typically employed with tissue culture-derived biologics to ensure the absence of mycoplasma, which is not a concern for plant-derived products. TMV passes readily through 0.45-μm membranes. However, the introduction of 0.2-μm filters was prompted by the identification of diminutive bacteria, that consistently penetrated 0.45-μm filters (Bowman et al. 1967). These pathogens are widely distributed in nature and have been isolated from water (Howard and Duberstein 1980; Sundaram et al. 1999). Exclusion of these pathogens from the buffers used during extraction, purification, and formulation can be achieved, but the field or greenhouse-cultivated infected plant material remains a potential source.

One approach for the sterilization of TMV fusion vaccines is to employ UV irradiation. UV treatment was applied to a canine parvovirus (CPV) vaccine, based on CPMV, and the vaccine preparation remained effective, protecting dogs from lethal challenge with CPV (Langeveld et al. 2001). When tested with a TMV fusion displaying a peptide from the VP2 capsid protein of CPV, the conditions required to inactivate bioburden resulted in cleavage of the coat protein, producing an ill-defined product that was difficult to qualify. As a result, we have not pursued this technique further. Other sources of ionizing radiation, e.g., gamma radiation, when combined with excipients to prevent protein aggregation (Assemand et al. 2003) and fragmentation (Moon and Bin Song 2001) may provide an alternative to UV treatment. Gamma irradiation is also an attractive route to sterilization, as it can be performed on the final vialed product.

Another sterilization methodology that may be applicable is the use of inactivating agents, such as formaldehyde, β-propiolactone and the aziridines. These agents are typically employed to inactivate viruses and bacterial toxins employed as vaccines, but they should prove equally effective at eliminating residual bioburden. Formaldehyde, which acts as a microbicide due to its peptide cross-linking activity, is employed in the manufacture of at least eight vaccines licensed for use in the United States (Offit and Jew 2003). Following treatment, the remaining free formaldehyde may be neutralized by the addition of sodium bisulfite (Martin et al. 2003). In spite of the widespread use of formaldehyde, it is not considered the ideal inactivant, as the inactivation reaction is neither linear nor first order (Wesslen et al. 1957) and extended inactivation periods are required (up to 60 days). Furthermore, since the nucleic acid remains functional, incomplete inactivation is possible, as has occurred with a foot-and mouth disease (FMD) vaccine and an early inactivated polio vaccine (Bahnemann 1990; Brown 2001). Consequently, inactivants that target nucleic acids may be more appropriate. One option is β-propiolactone (BPL), which functions by alkylating nucleic acids, thereby abolishing replication. It is currently employed in the inactivation of the cell culture-derived rabies virus vaccine (Perez and Paolazzi 1997) that is currently licensed in more than 20 countries and gained FDA approval in the United States in 1997 (Dreesen 1997). BPL is considered a possible carcinogen in humans, but it undergoes rapid hydrolysis in aqueous solution, and the breakdown products are nontoxic (Perrin and Morgeaux 1995). Of greater concern is the fact that BPL reacts with several amino acids, which could impair the immunogenicity of the coat protein-displayed epitopes, a point that has also been raised with regard to formaldehyde inactivation (Brown 2001).

A promising alternative is the use of aziridine compounds, such as binary ethylenimine (BEI), which also function through the alkylation of nucleic acids. Following virus inactivation, the residual BEI is hydrolyzed by the addition of sodium thiosulfate, which is itself innocuous. Although ethylenimines have been shown to react with proteins (Kasermann et al. 2001), their impact on epitope conformation and accessibility was substantially less than either BPL or formaldehyde treatment (Blackburn and Besselaar 1991). LSBC has evaluated BEI as a means of eliminating bioburden from investigational vaccine preparations. To date, approximately ten different TMV epitope fusions have been tested with the BEI procedure. We have determined conditions that effectively inactivate bioburden and in all cases the displayed epitopes retained antigenicity (Palmer et al., 2006). This procedure can be readily incorporated into veterinary vaccine production, since BEI inactivation is already used in the preparation of the current FMD vaccine, which is distributed globally and is the highest volume viral vaccine manufactured (Bahnemann 1990). BEI inactivation has not yet been approved for use with human biologics; however, no toxicity or tumorigenicity has been observed in the billions of livestock vaccinated with aziridine-inactivated products (Brown et al. 1998). In addition, a chemically related compound, PEN110, is currently in phase III clinical trials as a method of pathogen eradication in red blood cells, for patients requiring acute and chronic transfusions (Wu and Snyder 2003). For these reasons, the aziridines have been proposed as a universal method for pathogen inactivation in blood products, as well as in biopharmaceutical manufacture (Brown et al. 1998), and appear, from our experience, to be readily applicable to the preparation of TMV peptide vaccines. BEI treatment also eliminates TMV infectivity (LSBC, unpublished data). Inactivation of TMV may be important in developing these products for use in veterinary medicine, since state Departments of Agriculture regulate movement of infectious plant pathogens. We, and others, have shown that TMV is incapable of replication in mammalian cells, so elimination of infectivity is not a human or animal safety issue per se, but rather a plant health concern.

Once formulated, sterilized and vialed, the TMV fusion enters a stability study to ensure that protein integrity is maintained and to evaluate shelf life. For example, one ongoing stability study at LSBC is for a TMV fusion displaying a 13 amino acid peptide from the VP2 protein of canine parvovirus. This fusion constitutes a potential veterinary vaccine against canine parvovirus, feline panleukopenia virus and mink enteritis virus, as the neutralizing epitope is conserved across all three viruses (Dalsgaard et al. 1997). The vialed vaccine was stored as a 2-mg/ml solution at 4°C. Data for the 14-month time point is shown in Fig. 7a, b. Excellent long-term stability for this fusion was observed as evidenced by SDS-PAGE analysis and from MALDI-TOF, which confirmed the presence of the full epitope.

In another stability study for a series of investigational papillomavirus vaccines, consisting of epitopes from the L2 surface protein fused to TMV, the vialed products were stored at –20°C. Coat protein fusion purity, defined as the percentage of full-length product, was determined by SDS-PAGE, and MALDI-TOF analysis was performed to identify all species. The initial and 6-month data for four of the TMV fusions is summarized in Fig. 7c. Excellent stability was observed for L2 peptides #1 and #3, whereas cleavage occurred with the remaining two TMV fusions over the course of the study. This example highlights the need to completely characterize each coat protein fusion. Stability of the frozen products is also important since, during vaccine production, the purified biologic is often frozen prior to formulation to permit inventory build-up and for storage prior to formulation in a campaign mode (Patro et al. 2002). The cause for the observed

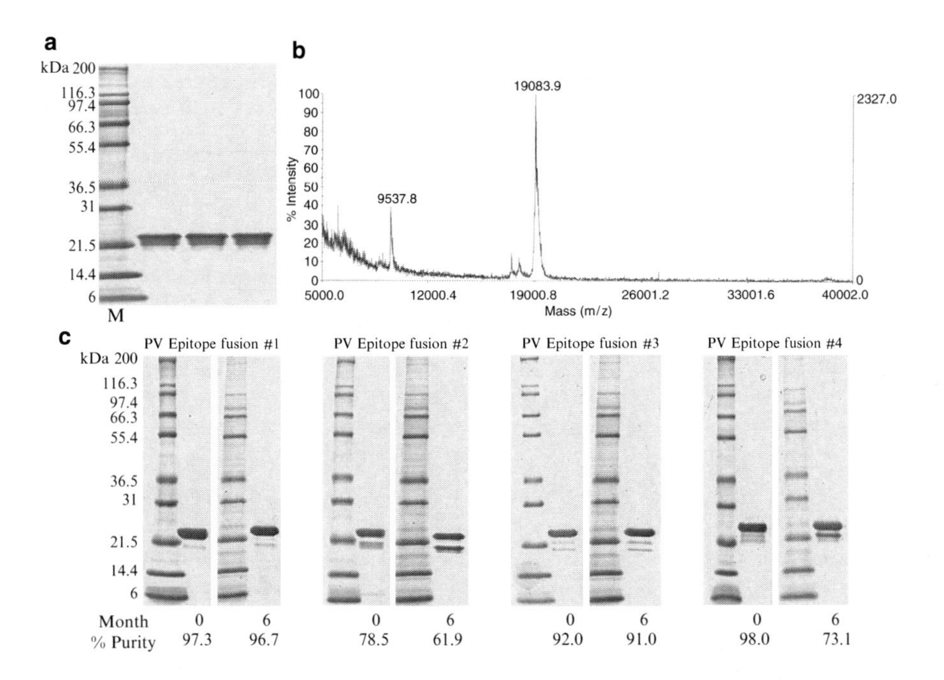

Fig. 7 a–c Stability of TMV fusion vaccines with storage. **a** Stability study for TMV fusion displaying a 13 amino acid peptide from the VP2 protein of canine parvovirus. Vialed vaccine, at 2 mg/ml, was stored for 14 months at 4°C. SDS-PAGE analysis of vaccine, analyzed in triplicate: purity estimated at more than 98%. **b** MALDI-TOF mass spectrometry spectrum for TMV fusion vaccine shown in **a**. The expected MW was 19,081 Da and the observed MW was 19,084 Da. **c** Stability study for four papillomavirus (*PV*) L2 peptide TMV fusions. SDS-PAGE analysis for the initial vialed vaccine and the vaccines after storage for 6 months at –20°C. Percent purity (full-length species) is indicated. For analysis each sample was run in triplicate. The identity of the coat protein fusions and any observed cleavage product was determined by mass spectrometry

reduction in full-length coat protein fusion for certain fusions is currently under investigation. One possibility is that the hydrolysis of the protein backbone is catalyzed by a host-derived protease, present in the purified vaccine preparation at below Coomassie-stainable levels (<0.5%). If this is the case, a polishing chromatography step or steps could potentially be employed to separate the proteolytic activity from the TMV fusion.

Summary

Many groups have developed vaccines based on peptide display on the surface of plant viruses that show exciting potential for use as prophylactics and therapeutics. However, the challenges associated with bringing new biological products into practical use in human and animal healthcare are underappreciated. Since plant virus peptide display vaccines have not yet been administered to humans, except in crude edible vaccine form, many of the manufacturing, quality control, and quality assurance methodologies necessary to bring the technology into more general use have not been described.

From a commercial perspective, development of vaccines that can prevent infectious diseases poses challenging business questions. Infectious disease poses the greatest risk to human health in poorer areas of the world, resulting in the requirement for low cost-of-goods, and lower profit margins than are traditionally attractive to larger companies. For a plant-based system such as the TMV peptide display platform, the exceptionally low cost-of-goods and outstanding environmental stability of TMV vaccines may make this a viable system for delivery of vaccines in resource-poor conditions. Data on safety and immunogenicity of these vaccines in humans are urgently required to move the product concept forward.

For companies developing new vaccine technologies, such as VLP epitope display platforms, it is also attractive to consider immunotherapeutic products that have lower development costs than prophylactic vaccines against infectious diseases, which require very large clinical trials over long periods of time. The concept of using vaccination to break immune tolerance to self-antigens and induce antibodies that may act therapeutically is a new and exciting one (Bachmann and Dyer 2004). It is well known that the best way to achieve this is through linkage of critical B cell epitopes to virus-like particles (Fehr et al. 1998; Chackerian et al. 2002). The report by Fitchen et al. (1995) provides data showing that TMV particles may be useful for the display of self-peptides in the development of therapeutic vaccines for induction of autoantibodies for the treatment of chronic diseases, such as Alzheimer's disease, rheumatoid arthritis, Crohn's disease, psoriasis, allergy, obesity, and drug addiction, and various infectious diseases, among others. These are drug targets of many vaccine and pharmaceutical companies. Therapeutic vaccines against chronic diseases also frequently require functional T cell immunity, directed against chronically infected or cancerous tissues, TMV displayed peptides do indeed promote functional T cell responses (McCormick et al. 2006). A fundamental requirement for therapeutic vaccines to be effective is that they activate professional APCs that can prime T cells reactive against target cells and tissues, and induce a strong Th1-biased immune response. The data published by Loor (1967) suggest that TMV particles are engulfed by APCs and rapidly transported to lymphoid organs for presentation to T cells. However, recent data suggest that different VLPs have very distinct properties when their abilities to activate the innate immune system are compared, and that the use of adjuvants, as well as elegant molecular engineering strategies can be applied to modulate the activity of epitope-display vaccines (Chackerian et al. 2002; Storni et al. 2004). Investigation of the interaction of plant virus particles with the innate immune system is an area of research that must be addressed in order to make plant virus-based epitope carriers competitive with other systems. When these data are coupled with industrial-scale methods for manufacturing, plant virus capsid-based epitope display systems could provide valuable contributions to human and veterinary health care in the near future.

Acknowledgements The authors thank Gregory Pogue, Stephen Garger, and Kathleen Hanley for insightful discussions and suggestions in the preparation of this manuscript. We also acknowledge the scientific staff at LSBC, who provided data for this review. In particular, we thank Tiffany Bliss, Amanda Lasnik, Alison McCormick, Michael McCulloch, Michael Mullican, Long Nguyen, and Sarah Venneman.

References

Adams SE, Dawson KM, Gull K, et al (1987) The expression of hybrid HIV-Ty virus-like particles in yeast. Nature 329:68–70

Assemand E, Lacroix M, Mateescu MA (2003) L-Tyrosine prevents aggregation of therapeutic proteins by gamma-irradiation. Biotechnol Appl Biochem 38:151–156

Bachmann MF, Dyer MR (2004) Opinion. Therapeutic vaccination for chronic diseases: a new class of drugs in sight. Nat Rev Drug Discov 3:81–88

Bahnemann HG (1990) Inactivation of viral-antigens for vaccine preparation with particular reference to the application of binary ethyleneimine. Vaccine 8:299–303

Bendahmane M, Koo M, Karrer E, et al (1999) Display of epitopes on the surface of tobacco mosaic virus: impact of charge and isoelectric point of the epitope on virus-host interactions. J Mol Biol 290:9–20

Birkett A, Lyons K, Schmidt A, et al (2002) A modified hepatitis B virus core particle containing multiple epitopes of the *Plasmodium falciparum* circumsporozoite protein provides a highly immunogenic malaria vaccine in preclinical analyses in rodent and primate hosts. Infect Immun 70:6860–6870

Blackburn NK, Besselaar TG (1991) A study of the effect of chemical inactivants on the epitopes of Rift-Valley fever virus glycoproteins using monoclonal-antibodies. J Virol Methods 33:367–374

Bowman F, Calhoun M, White M (1967) Microbiological methods for quality control of membrane filters. J Pharm Sci 56:222–225

Brown F (2001) Inactivation of viruses by aziridines. Vaccine 20:322–327

Brown F, Meyer RF, Law M, et al (1998) A universal virus inactivant for decontaminating blood and biopharmaceutical products. Biologicals 26:39–47

Burke KL, Dunn G, Ferguson Met al (1988) Antigen chimeras of poliovirus as potential new vaccines. Nature 332:81–82

Chackerian B, Lenz P, Lowy DR, et al (2002) Determinants of autoantibody induction by conjugated papillomavirus virus-like particles. J Immunol 169:6120–6126

Clarke BE, Newton SE, Carroll AR, et al (1987) Improved immunogenicity of a peptide epitope after fusion to hepatitis-B core protein. Nature 330:381–384

Clarke BE, Brown AL, Grace KG, et al (1990) Presentation and immunogenicity of viral epitopes on the surface of hybrid hepatitis-B virus core particles produced in bacteria. J Gen Virol 71:1109–1117

Corbett MK (1961) Purification of potato virus X without aggregation. Virology 15:8–15

Dalsgaard K, Uttenthal A, Jones TD, et al (1997) Plant-derived vaccine protects target animals against a viral disease. Nat Biotechnol 15:248–252

Dawson WO, Beck DL, Knorr DA, et al (1986) cDNA cloning of the complete genome of tobacco mosaic-virus and production of infectious transcripts. Proc Natl Acad Sci U S A 83:1832–1836

Delpeyroux F, Chenciner N, Lim A, et al (1986) A poliovirus neutralization epitope expressed on hybrid hepatitis-B surface-antigen particles. Science 233:472–475

Delpeyroux F, Peillon N, Blondel B, et al (1988) Presentation and immunogenicity of the hepatitis-B surface-antigen and a poliovirus neutralization antigen on mixed empty envelope particles. J Virol 62:1836–1839

Dreesen DW (1997) A global review of rabies vaccines for human use. Vaccine 15:S2–S6
Dunn DB, Hitchborn JH (1965) The use of bentonite in the purification of plant viruses. Virology 25:171–192
Fehr T, Skrastina D, Pumpens P, et al (1998) T cell-independent type I antibody response against B cell epitopes expressed repetitively on recombinant virus particles. Proc Natl Acad Sci U S A 95:9477–9481
Fitchen J, Beachy RN, Hein MB (1995) Plant-virus expressing hybrid coat protein with added murine epitope elicits autoantibody response. Vaccine 13:1051–1057
Fitzmaurice WP (2002) U.S. Patent No. 6344597
Garger SJ, Holtz RB, et al (2000) U.S. patent No. 6033895
Garger SJ, Holtz RB, et al (2001) U.S. patent No. 6303779
Gooding GV, Hebert TT (1967) A simple technique for purification of tobacco mosaic virus in large quantities. Phytopathology 57:1285
Haynes JR, Cunningham J, von Seefried A, et al (1986) Development of a genetically-engineered, candidate polio vaccine employing the self-assembling properties of the tobacco mosaic-virus coat protein. Bio/Technology 4:637–641
Howard G, Duberstein R (1980) A case of penetration of 0.2 um rated membrane filters by bacteria. J Parenter Drug Assoc 34:95
Kasermann F, Wyss K, Kempf C (2001) Virus inactivation and protein modifications by ethyleneimines. Antiviral Res 52:33–41
Koo M, Bendahmane M, Lettieri GA, et al (1999) Protective immunity against murine hepatitis virus (MHV) induced by intranasal or subcutaneous administration of hybrids of tobacco mosaic virus that carries an MHV epitope. Proc Natl Acad Sci U S A 96:7774–7779
Langeveld JPM, Brennan FR, Martínez-Torrecuadrada JL, et al (2001) Inactivated recombinant plant virus protects dogs from a lethal challenge with canine parvovirus. Vaccine 19:3661–3670
Loor F (1967) Comparative immunogenicities of tobacco mosaic virus, protein subunits, and reaggregated protein subunits. Virology 33:215–220
Marbrook J, Matthews REF (1966) The differential immunogenicity of plant viral protein and nucleoproteins. Virology 28:219–228
Martin A, Wychowski C, Couderc T, et al (1988) Engineering a poliovirus type-2 antigenic site on a type-1 capsid results in a chimaeric virus which is neurovirulent for mice. EMBO J 7:2839–2847
Martin J, Crossland G, Wood DJ, et al (2003) Characterization of formaldehyde-inactivated poliovirus preparations made from live-attenuated strains. J Gen Virol 84:1781–1788
McCormick AA, Reinl SJ, Cameron TI, et al (2003) Individualized human scFv vaccines produced in plants: humoral anti-idiotype responses in vaccinated mice confirm relevance to the tumor Ig. J Immunol Methods 278:95–104
McCormick AA, Corbo T, Wykoff-Clary S et al. (2006) TMV-peptide fusion vaccines induce T-cell mediated immune responses and tumor protection in two murine models. Vaccine 24:6414–6423
McCormick AA, Reddy S, Reinl SJ et al. (2008) Plant-produced idiotype vaccines for the treatment of Non-Hodgkin's Lymphoma: Safety and immunogenicity in a phase I clinical study. Proc. Natl. Acad. Sci. USA 105 (29):10131–10136
Meshi T, Ishikawa M, Motoyoshi F, et al (1986) In vitro transcription of infectious RNAs from full-length cDNAs of tobacco mosaic-virus. Proc Natl Acad Sci U S A 83:5043–5047
Moon S, Bin Song K (2001) Effect of gamma-irradiation on the molecular properties of ovalbumin and ovomucoid and protection by ascorbic acid. Food Chem 74:479–483
Moorthy VS, Good MF, Hill AV (2004) Malaria vaccine developments. Lancet 363:150–156
Namba K, Stubbs G (1986) Structure of tobacco mosaic-virus at 3.6-Å resolution—implications for assembly. Science 231:1401–1406
Offit PA, Jew PK (2003) Addressing parents' concerns: do vaccines contain harmful preservatives, adjuvants, additives, or residuals? Pediatrics 112:1394–1401
Palmer KE, Benko A Doucette SA et al. (2006) Protection of rabbits against cutaneous and mucosal papillomavirus infection using recombinant tobacco mosaic virus containing L2 capsid epitopes. Vaccine 24:5516–5525

Patro SY, Freund E, Chang BS (2002) Protein formulation and fill-finish operations. Biotechnol Annu Rev 8:55–84

Perez O, Paolazzi CC (1997) Production methods for rabies vaccine. J Ind Microbiol Biotechnol 18:340–347

Perrin P, Morgeaux S (1995) Inactivation of DNA by beta-propiolactone. Biologicals 23:207–211

Pogue GP, Lindbo JA, Garger SJ, et al (2002) Making an ally from an enemy: plant virology and the new agriculture. Annu Rev Phytopathol 40:45–74

Pogue GP, Lindbo JA, McCulloch MJ, et al (2004) U.S. patent No. 6730306 B1

Porta C, Spall VE, Findley KC, et al (2003) Cowpea mosaic virus-based chimaeras—effects of inserted peptides on the phenotype, host range, and transmissibility of the modified viruses. Virology 310:50–63

Powell MF, Nguyen T, Baloian L (1998) Compendium of excipients for parenteral formulations. PDA J Pharm Sci Technol 52:238–311

Staczek J, Bendahmane M, Gilleland LB, et al (2000) Immunization with a chimeric tobacco mosaic virus containing an epitope of outer membrane protein F of *Pseudomonas aeruginosa* provides protection against challenge with *P. aeruginosa*. Vaccine 18:2266–2274

Storni T, Ruedl C, Schwartz K, et al (2004) Nonmethylated CG motifs packaged into virus-like particles induce protective cytotoxic T cell responses in the absence of systemic side effects. J Immunol 172:1777–1785

Sugiyama Y, Hamamoto H, Takemoto S, et al (1995) Systemic production of foreign peptides on the particle surface of tobacco mosaic-virus. FEBS Lett 359:247–250

Sundaram S, Auriemma M, Howard G Jr, et al (1999) Application of membrane filtration for removal of diminutive bioburden organisms in pharmaceutical products and processes. PDA J Pharm Sci Technol 53:186–201

Timian RG, Savage SM (1966) Purification of barley stripe mosaic virus with chloroform and charcoal. Phytopathology 56:1233–1235

Turpen TH (1999) Tobacco mosaic virus and the virescence of biotechnology. Philos Trans R Soc Lond B Biol Sci 354:665–673

Turpen TH, Reinl SJ, Charoenvit Y, et al (1995) Malarial epitopes expressed on the surface of recombinant tobacco virus. Bio/Technology 13:53–57

Valenzuela P, Coit D, Kuo G (1985) Antigen engineering in yeast—synthesis and assembly of hybrid hepatitis-B surface antigen herpes simplex 1 gD particles. Bio/Technology 3:323–326

van Regenmortel MHV (1999) The antigenicity of tobacco mosaic virus. Philos Trans R Soc Lond B Biol Sci 354:559–568

Watson JD (1954) The structure of tobacco mosaic virus 1. X-ray evidence of a helical arrangement of sub-units around the longitudinal axis. Biochim Biophys Acta 13:10–19

Wesslen T, Lycke E, Olin G, et al (1957) Inactivation of poliomyelitis virus by formaldehyde. Arch Gesamte Virusforsch 7:125–135

Wu LG, Jiang LB, Zhou Z, et al (2003) Expression of foot-and-mouth disease virus epitopes in tobacco by a tobacco mosaic virus-based vector. Vaccine 21:4390–4398

Wu YY, Snyder EL (2003) Safety of the blood supply: role of pathogen reduction. Blood Rev 17:111–122

Yusibov V, Hooper DC, Spitsin SV, et al (2002) Expression in plants and immunogenicity of plant virus-based experimental rabies vaccine. Vaccine 20:3155–3164

Chloroplast-Derived Vaccine Antigens and Biopharmaceuticals: Expression, Folding, Assembly and Functionality

S. Chebolu and H. Daniell

Contents

Abstract Chloroplast genetic engineering offers several advantages, including high levels of transgene expression, transgene containment via maternal inheritance, and multi-gene expression in a single transformation event. Oral delivery is facilitated by hyperexpression of vaccine antigens against cholera, tetanus, anthrax, plague, or canine parvovirus (4%–31% of total soluble protein, TSP) in transgenic chloroplasts (leaves) or non-green plastids (carrots, tomato) as well as the availability of antibiotic free selectable markers or the ability to excise selectable marker genes. Hyperexpression of several therapeutic proteins, including human serum albumin (11.1% TSP), somatotropin (7% TSP), interferon-alpha (19% TSP), interferon-gamma (6% TSP), and antimicrobial peptide (21.5% TSP), facilitates efficient and economic purification. Also, the presence of chaperones and enzymes in chloroplasts facilitates assembly of complex multisubunit proteins and correct folding of human blood proteins with proper disulfide bonds. Functionality of chloroplast-derived vaccine antigens and therapeutic proteins has been demonstrated by several assays, including the macrophage lysis assay, GM1-ganglioside binding assay, protection of HeLA cells or human lung

S. Chebolu and H. Daniell(✉)
Department of Molecular Biology and Microbiology, University of Central Florida, Biomolecular Science (Bldg. 20), Room 336 Orlando, FL 32816-2364, USA
e-mail: daniell@mail.ucf.edu

A.V. Karasev (ed.) *Plant-produced Microbial Vaccines.*
Current Topics in Microbiology and Immunology 332

carcinoma cells against encephalomyocarditis virus, systemic immune response, protection against pathogen challenge, and growth or inhibition of cell cultures. Purification of human proinsulin has been achieved using novel purification strategies (inverse temperature transition property) that do not require expensive column chromatography techniques. Thus, transgenic chloroplasts are ideal bioreactors for production of functional human and animal therapeutic proteins in an environmentally friendly manner.

Introduction

Vaccines and therapeutic proteins are the great successes of modern medicine, which have been used for several decades to prevent diseases and eradicate them. The uses of vaccines and therapeutic proteins have great potential but are limited by their cost of production, distribution, and delivery. Modified mammalian cells are used for producing therapeutic proteins, which have the advantage of resulting in products that are similar to their natural counterparts. These cells can be cultured on a limited scale but production is quite expensive. Bacteria can be used for large-scale production of proteins, but the products differ from the natural products considerably. For example, the proteins that are usually glycosylated in humans are not glycosylated by bacteria. Moreover, many proteins are expressed in *Escherichia coli* on a large scale, but sometimes they may differ in conformation with eventual precipitation due to a lack of proper folding and disulfide bridges (Daniell et al. 2001d).

For ages, humans have been using plants as a source of food, clothing, medicine, and building materials. Plants have been of immense help in the past and continue to be so. Plants are now one of the new hosts for the production of biopharmaceuticals, polymers, vaccines, enzymes, plasma proteins, and antibodies. There are many advantages in production of recombinant proteins in plants. Primarily, plant systems are more economical in that they can be produced on a large scale rather than using industrial methods (fermentation of bacteria, yeast or cultured animal or human cell lines) that are very expensive. Also, there is no need to maintain the cold chain as the plant parts expressing the vaccine or plant extracts can be stored and transported at room temperature. Plants have the ability to carry out posttranslational modifications similar to naturally occurring systems. There is also minimized risk of contamination from potential human pathogens, as plants are not hosts for human infectious agents (Giddings et al. 2000). If therapeutic proteins are delivered orally, then the purification step from plants can be eliminated. Other advantages of food-based vaccines include the low cost of raw material and convenient storage; the cost of syringes and needles in delivery of vaccines is eliminated, which thereby eliminates blood-borne pathogens.

Plant Expression System

Many different proteins with applications for human or animal vaccines, biopharmaceuticals, and biopolymers have been expressed in transgenic plants. With a few exceptions, most often very low expression levels of foreign proteins (less than 1% of the total soluble protein) were observed in nuclear transgenic plants (Daniell et al. 2001d), including the B subunit of enterotoxigenic *E. coli* (0.01% TSP; Haq et al. 1995), hepatitis B virus envelope surface protein (0.01% TSP; Mason et al. 1992; Thanavala et al. 1995), human cytomegalovirus glycoprotein B (0.02% TSP; Tackaberry et al. 1999), and transmissible gastroenteritis coronavirus glycoprotein S (0.06% TSP; Gomez et al. 1998). Also, gene silencing can occur in nuclear transformation, which results in lower expression of recombinant proteins. For commercial exploitation of the therapeutic proteins and vaccine antigens, high and reliable levels of expression are required, which could be achieved by alternative approaches.

Chloroplast Expression System

The highest level of protein expression of *Bacillus thuringiensis* (Bt) cry2Aa2 in plants was achieved at 46.1% TSP in transgenic tobacco chloroplasts (De Cosa et al. 2001). Besides such high expression levels, there are several other advantages of chloroplast genetic engineering over nuclear transformation. Chloroplast genomes defy the laws of Mendelian inheritance in that they are maternally inherited in most crops (Zhang et al. 2003) and thus minimize out-crossing of transgenic pollen with related weeds or crops. Even if the pollen from plants that exhibit maternal inheritance contains metabolically active plastids, the plastid DNA is lost during pollen maturation and is not transmitted to the next generation (Daniell 2002). Therefore, the chloroplast expression system is an environmentally friendly approach. Also, chloroplasts have the ability to express multiple genes in a single transformation event. Expression of polycistrons in transgenic chloroplasts is a unique feature, which facilitates the expression of entire pathways in a single transformation event (De Cosa et al. 2001; Daniell and Dhingra 2002). For the first time, a complete bacterial operon was successfully expressed in transgenic chloroplasts, resulting in the formation of stable cry2Aa2 crystals (De Cosa et al. 2001). This should facilitate expression of polyvalent vaccines or multisubunit proteins in transgenic chloroplasts.

In addition, the position effect seen in nuclear transgenics can be eliminated as the transgenes are targeted into spacer regions at precise locations in the chloroplast genome by homologous recombination of chloroplast DNA flanking sequences. The problem of gene silencing has not been observed in transgenic chloroplasts in spite of higher expression levels than in nuclear transgenic plants. It has been observed that there is no gene silencing in spite of high translation levels, up to

46.1% TSP (De Cosa et al. 2001) or transcription in transgenic chloroplasts, despite accumulation of transcripts 169-fold and 150-fold higher than nuclear transgenics (Lee et al.2003; Dhingra et al. 2004).

Chloroplasts also have the ability to process eukaryotic proteins, including correct folding of subunits and formation of disulfide bridges (Daniell et al. 2001d). Functional assays showed that chloroplast-synthesized cholera toxin-B subunit binds to the intestinal membrane GM1-ganglioside receptor, thereby confirming the correct folding and disulfide bond formation (Staub et al. 2000; Daniell et al. 2001b; Molina et al. 2004). Chaperones present in chloroplasts facilitate correct folding and assembly of monoclonal antibody in transgenic chloroplasts (Daniell et al. 2004a) and also result in fully functional human therapeutic proteins, as seen in interferon alpha and gamma (Falconer 2002; Leelavathi and Reddy 2003). Chloroplasts can be a good place to store the biosynthetic products that could otherwise be harmful when accumulated in cytosol (Bogorad 2000). This was demonstrated when cholera toxin B subunit was accumulated in large quantities in transgenic chloroplasts and it had no toxic effect (Daniell et al. 2001b), whereas when accumulated in the cytosol in very small quantities, CTB was toxic (Mason et al. 1998). Similarly, trehalose, which is used as a preservative in the pharmaceutical industry, was toxic when accumulated in cytosol but was nontoxic when compartmentalized within chloroplasts (Lee et al. 2003). Several therapeutic proteins and vaccine antigens expressed via the chloroplast genome are listed in Tables 1 and 2.

Novel Purification Strategies

The main reason for the high cost of pharmaceutical protein production is purification of recombinant proteins. Therefore, novel protein purification strategies can be used that do not require the use of expensive column chromatography. For instance, a synthetic protein-based polymer gene $(GVGVP)_{121}$ has been expressed in *E. coli* and a very high expression level was achieved such that polymer inclusion bodies were formed that occupied nearly 90% of the cell volume (Daniell et al. 1997; Guda et al. 2000). $(GVGVP)_{121}$ exhibits inverse temperature transition properties, making it soluble in water below room temperature, but aggregates into a more ordered, viscoelastic state, called a coacervate, at 37°C (Daniell et al. 1997; Guda et al. 2000). The inverse temperature transition property makes purification easier and less expensive in aqueous solutions simply by raising the temperature. In addition, this property makes $(GVGVP)_{121}$ an ideal fusion protein for purification. To demonstrate this, the proinsulin gene was fused to a smaller version of this biopolymer (GVGVP) 40 with inverse temperature transition properties and was expressed in *E. coli* (Daniell et al. 2004a). At 4°C, the biopolymer exists as an extended molecule, but when incubated at 42°C, it folds into dynamic structures called β-spirals that further aggregate by hydrophobic association to form twisted filaments (Urry et al. 1996). Therefore, this principle was successfully used to

Table 1 List of vaccine antigen proteins expressed via the chloroplast genome

Vaccine antigens	Gene	Site of integration	Promoter	5'/3' Regulatory elements	% TSP expression	Functionality assay	Lab/year
Cholera toxin	*Ctx*B	*Trn*I/*trn*A	Prrn	ggagg/ T*psb*A	4%	GM–1 ganglioside binding assay	Daniell 2001
Tetanus toxin	*Tet*C (bacterial and synthetic)	*Trn*V/*rps*12/7	Prrn	T7 gene 10a, *atp*B[b] / T*rbc*L	25% [a], 10% [b]	Positive systemic immune response	Maliga 2003
Canine parvovirus (CPV)	CTB–2L21/GFP–2L21	*Trn*I/*trn*A	Prrn	P*psb*A/T*psb*A	31.1%, 22.6%	Immune response	Daniell, Veramandi 2004
Anthrax protective antigen	*Pag*	*Trn*I/*trn*A	Prrn	P*psb*A/T*psb*A	4%–5%	Macrophage lysis assay	Daniell 2004
Plague vaccine	*Ca*F1~*Lcr*V	*Trn*I/*trn*A	Prrn	P*psb*A/T*psb*A	14.8%	ND	Daniell 2003

Table 2 List of biopharmaceutical proteins expressed via the chloroplast genome

Biopharmaceutical proteins	Gene	Site of Integration	Promoter	5'/3' Regulatory elements	% TSP expression	Functionality assay	Lab, year
Elastin-derived polymer	EG121	*trn*I/*trn*A	Prrn	T7gene10/T*psb*A	ND	Inverse temp. transition assay	Daniell 2000
Human somatotropin	*hST*	*trn*V/*rps*12/7	Prrn[a], P *psb*A[b]	T7gene10[a] or *psb*A[b]/ T*rps*16	7.0% [a] and 1.0% [b]	Positive growth response of Nb2 cell line	Monsanto 2000
Antimicrobial peptide	MSI–99v	*trn*I/*trn*A	Prrn	ggagg/T*psb*A	21.5%–43%	Minimum inhibitory conc. (MIC) of *P. aeruginosa*	Daniell 2001
Insulin-like growth factor	*IGF*–1	*trn*I/*trn*A	Prrn	P*psb*A/ T*psb*A	33%	ND	Daniell 2002
Interferon-α 5	*INF*α 5	*trn*I/*trn*A	Prrn	P*psb*A/T*psb*A	ND	ND	Daniell 2002
Interferon-α 2b	*INF*α 2B	*trn*I/*trn*A	Prrn	P*psb*A/T*psb*A	18.8%	Protection of HeLA cells against cytopathic effects of EMC virus	Daniell 2002
Human serum albumin	*hsa*	*trn*I/*trn*A	Prrn[a], P*psb*A[b]	ggagg[a], *psb*A[b]/T*psb*A	0.02%[a], 11.1% [b]	ND	Daniell 2003
Interferon-γ	*IFN-g*	*rbc*L/*acc*D	P*psb*A	P*psb*A/T*psb*A	6%	Protection of human lung carcinoma cells against infection by EMC virus	Reddy 2003
Monoclonal antibodies	*Guy's 13*	*trn*I/*trn*A	Prrn	ggagg/ T*psb*A	ND	ND	Daniell 2001

purify the fusion protein simply by raising the temperature to 42°C and was analyzed on SDS-polyacrylamide gel. The presence of polymer in the proinsulin-polymer fusion protein was confirmed by negatively staining the SDS-PAGE gels with 0.3 M Cucl2 (Fig. 1a). Copper stained gels, when illuminated obliquely, show dark bands against a light, semiopaque background. The same gel was stained with Coomassie R-250. The sulphonic acid groups in the dye (Coomassie R-250) form ion pairs with lysine and arginine in the protein, which are not present in the polymer but are present in proinsulin. Since the Coomassie R-250 stained the polypeptide, it confirms the presence of fusion protein (Fig. 1b). After purification, the fusion protein can be cleaved with suitable enzymes and the polymer can be discarded in the pellet fraction by another round of polymerization at 42°C followed

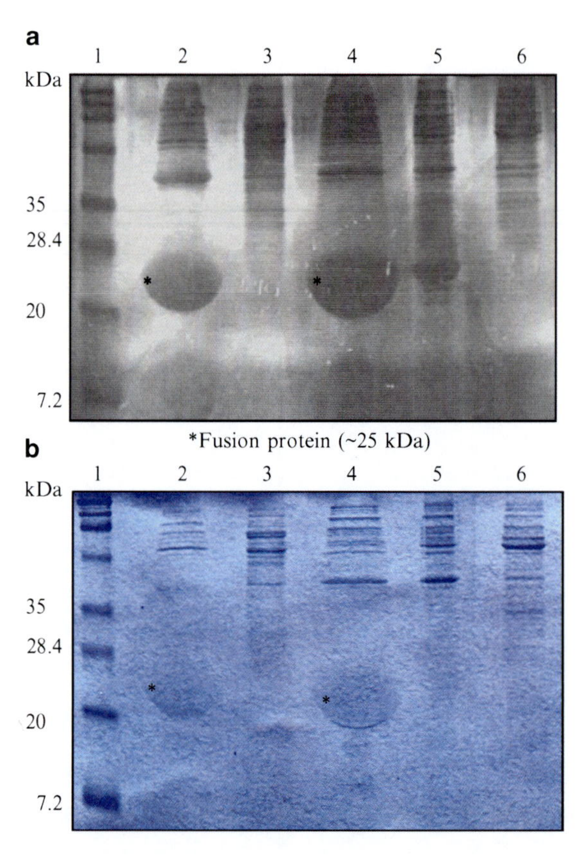

Fig. 1 a, b Expression and purification of insulin-polymer fusion protein detected in copper- (**a**) and Coomassie- (**b**) stained gels. Two rounds of purification through thermally reversible phase transition were performed. The same gel was first stained with copper, destained, and restained with Coomassie R-250. Lane 1: Prestained marker; lane 2: purified extract of polymer-insulin fusion protein from the chloroplast vector pSBL-OC-40Pris; lane 3: reverse orientation of fusion protein from pSBL-OC-40Pris; lane 4: purified extract of pLD-OC-40Pris; lane 5: reverse orientation of pLD-OC-40Pris; lane 6: purified extract of *E. coli* strain XL-1 blue containing no plasmid

by centrifugation. This novel purification strategy should significantly decrease the purification cost because of the elimination of column chromatography.

Oral Delivery of Vaccine Antigens

Oral delivery of vaccines is an attractive alternative because of the ease of administration and low costs. Also, oral delivery of the recombinant protein could reduce the production costs by almost 90%. As pointed out above, chloroplast genetic engineering is most suitable for hyperexpression of vaccine antigens and therapeutic proteins. However, oral delivery of vaccine antigens could significantly reduce the cost of their production, purification, storage, and transportation, thereby eliminating the need to maintain the cold chain. To achieve oral delivery of vaccines, one of the important requirements is the production of therapeutic proteins in antibiotic free selection.

Most transformation techniques use an antibiotic resistance gene along with the gene of interest such that only the transgenic lines containing the desired trait are selected. But once the trait has been established, the antibiotic resistance gene has only one purpose: producing its products. One major concern is horizontal transfer of the antibiotic resistance genes to other related or unrelated organisms. However, now chloroplasts can be genetically modified without the use of antibiotic resistance genes. There are several ways of doing this. For example, the spinach betaine aldehyde dehydrogenase (BADH) gene has been developed as a selectable marker to transform the chloroplast genome (Daniell et al. 2001c, e). The toxic betaine aldehyde (BA) is converted to nontoxic glycine betaine by the chloroplast BADH enzyme. This glycine betaine also serves as an osmoprotectant and confers salt tolerance (Kumar et al. 2004). The BADH enzyme is present in only a few plant species that are adapted to salty and dry environments.

In addition, the selectable marker genes can be removed using short, 174-bp DNA repeats, but the gene of interest could be retained outside the direct repeats, thereby producing marker-free plants (Iatham and Day 2000). Another method of eliminating the selectable marker is by employing a transiently co-integrated vector with a single homologous flank (Klaus et al. 2004). The co-integrates are lost rapidly by recombination when the selection pressure is removed.

For oral delivery of vaccines, another important requirement is the expression of the vaccine antigens in plastids of non-green tissues. For the first time, stable and highly efficient plastid transformation of carrot using non-green tissues as explants, regenerated via somatic embryogenesis, has been reported (Kumar et al. 2004). A useful plant trait (salt tolerance) has been expressed for the first time in a non-solanaceous crop via the chloroplast genome. The carrot-specific plastid transformation vector pDD-Dc-aadA/BADH was constructed and carrot callus was bombarded. The bombarded callus was plated on selection medium containing 150 mg/l spectinomycin to obtain spectinomycin-resistant transgenic lines. The BADH enzyme activity was assayed in crude extracts of protein from both transformed and untransformed carrot cell cultures. BADH enzyme in the presence of betaine aldehyde converts NAD+ to NADH and the reaction rate was measured by an increase in absorbance at 340 nm due to the reduction of NAD+(Kumar et al. 2004). Crude extracts from transgenic plastids showed higher levels of BADH activity when compared to untransformed tissue (Fig. 2a). High BADH activity was observed in leaves, tap roots, and cells in suspension culture but the difference in activity may be due to variation in plastid genome copy numbers. The high BADH activity in carrot taproot may be due to a large number of chromoplasts, which is evident by the orange color (Fig. 2b). Also, the protein expression in carrot cells, root, and shoot was tested using Western blot analysis (Fig. 2c). These results were consistent with the BADH enzyme activity. Also, glycine betaine is found in many higher plants, bacteria, and mammalian animals, which they accumulate under conditions of water or salt stress and acts as an osmoprotectant (Figueroa-Soto et al. 1999). In plants and bacteria, glycine betaine can build up the cytoplasmic osmotic strength without preventing any cellular functions. BADH was reported to be present in porcine kidneys localized in both cortex and medulla (Figueroa-Soto et al. 1999).

Carrot is a biennial plant with a vegetative phase in the 1st year and the reproductive phase in the 2nd year. Harvesting the crop containing the transgene at the end of 1st year prevents gene flow through seeds or pollen because there is no reproductive system. Also, maternal inheritance of carrot chloroplast genomes adds to the environmentally friendly approach (Vivek et al. 1999). Somatic embryos of carrot are single-cell-derived and multiply through recurrent embryogenesis, which provides a uniform source of cell culture and a homogeneous

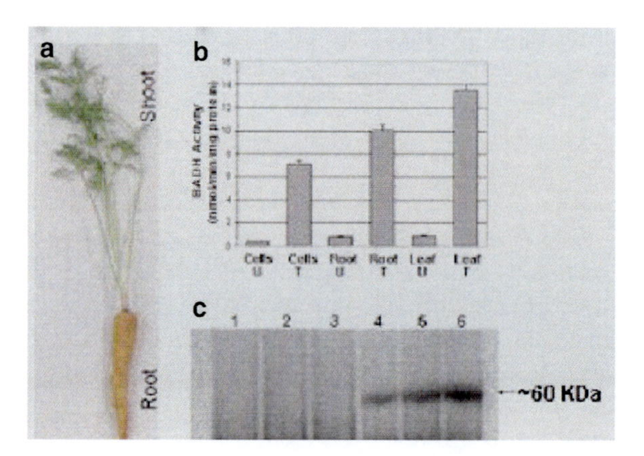

Fig. 2 **a–c** BADH enzyme activity and BADH expression in control and pDD-*Dc-aadA/BADH* lines. **a** BADH activity in untransformed (**U**) and transformed (**T**) cell suspension, root and leaf. **b** A pDD-*Dc-aadA/BADH* transgenic line shown with taproot and shoot. **c** Western blot using polyclonal anti-BADH serum. Antigenic peptides were detected using horseradish peroxidase-linked secondary antibody. Lanes 1, 2, 3, untransformed cell culture, root and leaf and lanes 4, 5, 6, transformed cell culture, root and leaf

single source, which is a requirement for the production of human therapeutic proteins. Oral delivery of therapeutic proteins via edible carrots preserves the structural integrity of the proteins, as no cooking is needed. Encapsulated embryos, which are viable for many years in culture, can be used for cryopreservation. Because of all these advantages, carrot can be used as an ideal system for oral delivery.

Chloroplast Derived Vaccine Antigens

Cholera Toxin B Antigen

Cholera toxin B subunit (CTB) of *Vibrio cholerae*, a candidate vaccine antigen, has been expressed in transgenic chloroplasts, and this resulted in accumulation of up to 4.1% total soluble protein as functional oligomers (Daniell et al. 2001b). Higher expression levels (up to 31.1% TSP) were obtained when CTB-2L21 fusion protein was expressed in transgenic chloroplasts (Molina et al. 2004). The difference in the expression levels of the CTB gene was due to the presence of a ribosome-binding site (GGAGG, 4.1% TSP) or the 5′ UTR of the psbA gene (31.1% TSP). CTB synthesized from transgenic chloroplasts in both investigations assembled into functional oligomers and were antigenically identical to purified native CTB. CTB has the highest affinity to bind with gangliosides (GM1), which are the natural toxin receptors in the intestinal epithelial cells. The functionality of CTB was determined by the GM1-ganglioside ELISA binding assay (Fig. 3), where the plates were coated first with GM1-gangliosides and BSA, which was then plated with the total soluble protein from the transformed and the untransformed plants and bacterial CTB. Then the absorbance of the GM1–ganglioside–CTB-antibody complex was measured. The bacterial CTB and chloroplast-synthesized CTB showed a strong affinity for the G_{M1}–ganglioside complex, confirming that the antigenic sites necessary for the binding of CTB to GM1 were conserved. With such high levels of expression of an efficient transmucosal carrier molecule such as CTB in chloroplasts, fusion proteins can be synthesized and also plant-derived vaccines can be commercialized.

Anthrax Protective Vaccine Antigen

The Center for Disease Control (CDC) lists *Bacillus anthracis* as a category A agent and estimates the cost of an anthrax attack to exceed $26 billion per 100,000 exposed individuals (Kaufmann et al. 1997). The present vaccine for anthrax is produced by cell-free filtrate of toxigenic, nonencapsulated strain of *B. anthracis* (Baillie 2001). In addition to the protective antigen (PA), the immunogenic portion, trace amounts of lethal factor and edema factor are present in the filtrate, which are considered to be toxic and cause side effects (Ivins et al.

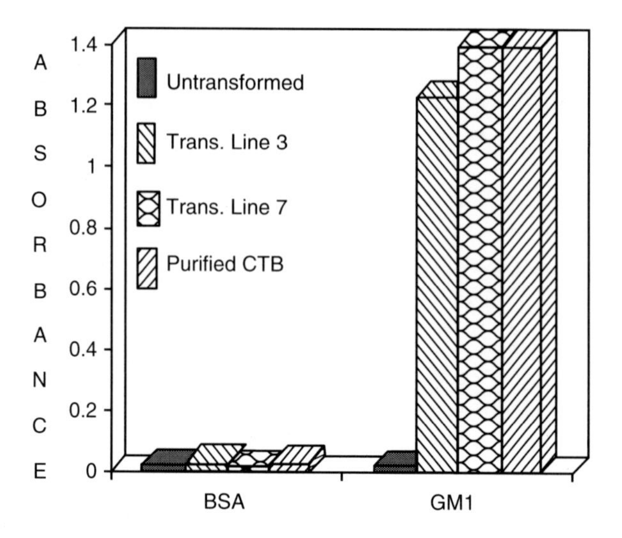

Fig. 3 CTB-GM1-ganglioside binding ELISA assay. Plates, coated first with GM1-ganglioside and bovine serum albumin (BSA), respectively, were irrigated with total soluble plant protein from chloroplast transgenic lines (3, 7) and 300 ng of purified bacterial CTB. The absorbance of the GM1-ganglioside-CTB-antibody complex in each case was measured at 405 nm. Total soluble protein from untransformed plants was used as the negative control

1995; Baillie 2001; Joellenback et al. 2002). The PA heptamerizes and binds itself to the host cell receptor. This can now bind to lethal factor or edema factor. When bound to lethal factor, the toxin stimulates the release of interleukins-1β by macrophages and other cytokines that lead to sudden death. For the above reasons and also because of the limited supply of the current vaccine, biothrax (formerly anthrax vaccine adsorbed, or AVA), there is an urgent need for an improved and pure vaccine. Therefore, the *B. anthracis* 83-kDa PA was expressed in transgenic tobacco chloroplasts. The PA gene was cloned into the chloroplast transformation vector (pLD-JW1) along with *psbA* regulatory signals to enhance translation. The crude plant extracts contained up to 2.5 mg PA/g fresh weight (Daniell et al. 2004b). The PA protein as a percentage of total soluble protein expressed in pLD-JW1 plants was determined for 3 and 5 days of continuous illumination. A maximum 18.1% and 13.4% PA in the total soluble protein was achieved (Fig. 4). The maximum levels of expression were observed in mature leaves under continuous illumination. Also, Western blots were performed using two different detergents in the extraction buffer, CHAPS and SDS, and both extracted PA equally well. After storage for 2 days at 4°C and –20°C, the PA in the crude extracts was quite stable. Powdered leaf was stored at –80°C for several months and then Western blots were performed, showing no noticeable decrease in PA quantity or functionality. This should facilitate the long-term storage of harvested leaves before PA is extracted for vaccine production. Trypsin, chymotrypsin, and furin proteolytic sites present in PA expressed in transgenic chloroplasts were protected in that only PA 83 was observed. The functionality of the antigen was determined by its ability to internalize the lethal factor and cause the lysis of cultured macrophage cells, which is termed as the macrophage lysis assay (Fig. 5). The percentage macrophage viability was deter-

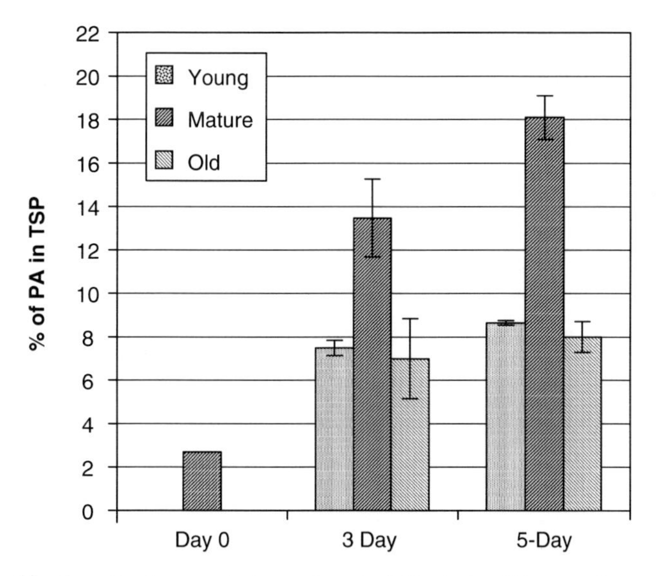

Fig. 4 Quantification of anthrax protective antigen in JW1 plants of T_1 generation using ELISA. Histogram of percentage of PA in the total soluble protein in fresh tissue of young, mature, and old leaves of plant in 16 h light and 8 h dark (day 0), 3-day continuous illumination and 5-day continuous illumination

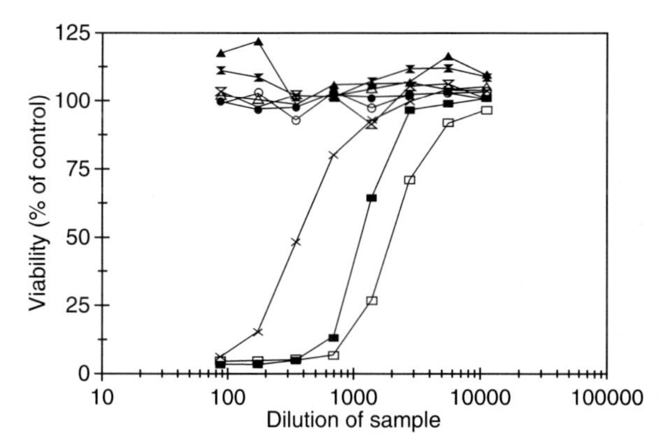

Fig. 5 Macrophage cytotoxic assays for extracts from transgenic plants. Supernatant samples from T1 pLD-JW1 tested (proteins extracted in buffer containing no detergent and MTT added after 5 h). pLD-JW1 (extract stored 2 days); pLD-JW1 (extract stored 7 days); PA 5 µg/ml; control wild type (extract stored 2 days); control wild type (extract stored 7 days); control wild type no LF (extract stored 2 days); control wild type no LF (extract stored 7 days); control pLD-JW1 no LF (extract stored 2 days); control pLD-JW1 no LF (extract stored 7 days)

mined using the ability of live but not dead cells to reduce a water-soluble yellow dye, MTT (3-[4,5-dimethylthiazol-2-yl [-2,5-diphenyltetrazolium bromide) to an insoluble purple formazan product (Daniell et al. 2004b). The PA also showed heptamerization and cleaved properly. With an average yield of 172 mg of PA per plant using the experimental transgenic cultivar grown in greenhouse, 400 million doses of vaccine (free of contaminants) could be produced per acre. The yield could increase further by 18-fold when a commercial cultivar is used rather than the experimental cultivar.

Plague Vaccine Antigen

Yersinia pestis, a Gram-negative bacterium, is the causative agent of plague and has been listed by Center for Disease Control (CDC 2003) as one of the six category A biological agents. There are three different forms of plague. Bubonic plague, the most common form of plague, is caused by infected fleas. When it enters the lymph nodes, the bacterium causes swelling of the nodes and forms buboes; the other forms are septicemic plague and pneumatic plague. The current vaccine available is a killed whole subunit vaccine, which is moderately effective against bubonic plague and ineffective against the other two forms of the plague (Titball and Williamson 2001). Several subunit vaccines have been evaluated for immunogenicity against *Y. pestis*. CaF1 and LcrV are the most effective subunit vaccines so far against *Y. pestis*. F1 is a capsular protein located on the surface of the bacterium with antiphagocytic properties. The V antigen is a component of the *Y. pestis* type III secretion system and it may form part of an injectosome. The fusion protein of F1-V expressed in *E. coli* has been shown to be safe and immunogenic when mice were challenged with *Y. pestis* (Williamson et al., 1997). The fusion protein F1-V was expressed in transgenic chloroplasts consisting of the F1 antigen fused at its carboxy terminus to the amino terminus of the V antigen. Western blot and ELISA were performed and samples were collected from plants under continuous illumination from days 0–5 from young, mature, and old leaves (Fig. 6). The maximum expression levels were observed in mature leaves, which is as high as 14.8% of the total soluble protein (Arlen et al. 2008). Further studies need to be done to confirm immunogenicity. Since *Y. pestis* is one of the category A biological warfare agents and there is no proper vaccine, improved vaccine is a high priority.

Canine Parvovirus VP2 Antigen

Canine parvovirus (CPV) infects dogs and other Canidae such as wolves, South American dogs, and Asiatic raccoon dogs, producing hemorrhagic gastroenteritis and myocarditis (CPV vaccination, Cornell University) .The 2L21 synthetic peptide, coupled to KLH carrier protein, was studied extensively and has been shown to be effective in protecting dogs and minks against parvovirus (Langeveld et al.

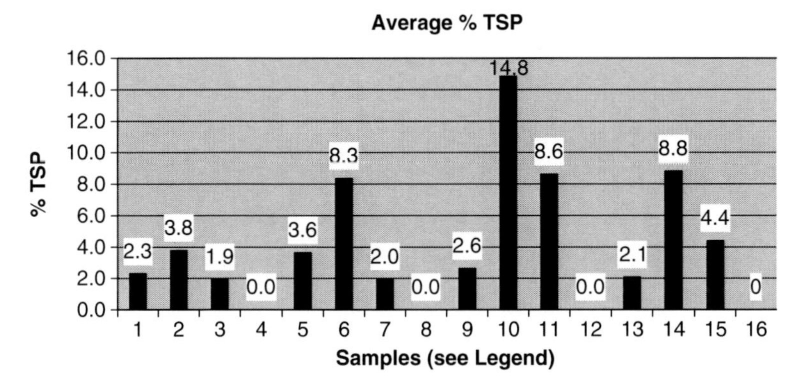

Fig. 6 Enzyme linked immunoassays expressed, with protein yields expressed as % TSP and micrograms of F1 V per gram of fresh leaf material. For the continuous illumination experiment, leaf material was sampled on days 0, 1, 3, and 5. Young, mature, and old samples were taken for each experiment. The figure shows average levels of F1 V antigen for samples 1–4 [day 0: (*1*) young, (*2*) mature, (*3*) old, and (*4*) wild-type leaf samples]; samples 5–8 [day 1: (*5*) young, (*6*) mature, (*7*) old and (*8*) wild-type leaf samples]; samples 9–12 [day 3: (*9*) young, (*10*) mature, (*11*) old and (*12*) wild-type leaf samples]; samples 13–16 [day 5: (*13*) young, (*14*) mature, (*15*) old, (*16*) wild-type leaf samples]

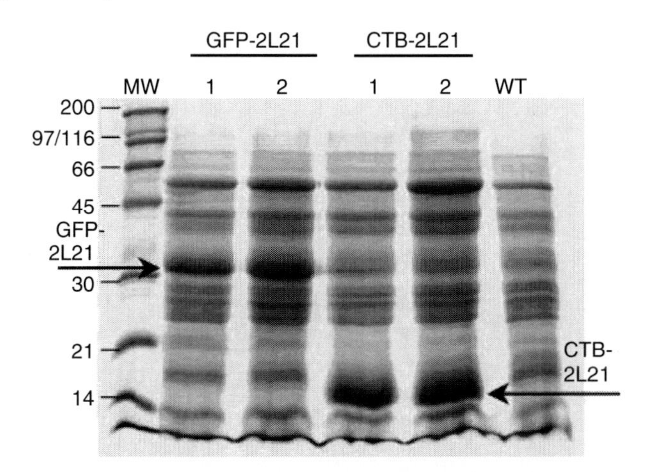

Fig. 7 Coomassie blue-stained sodium dodecylsulphate-polyacrylamide gel electrophoresis (SDS-PAGE) gel of plant samples, 60 days after transplanting. Two independent lines (1, 2) for each construction were analyzed. Recombinant proteins GFP-2L21 and CTB-2L21 are marked. Fifty micrograms of plant total protein was loaded per well. *2L21* epitope from the VP2 protein of the canine parvovirus, *CTB* cholera toxin B, *GFP* green fluorescent protein, *MW* molecular weight marker, *WT* wild-type Petit Havana plant

1994, 1995). The 2L21 peptide, which confers protection to dogs against virulent canine parvovirus (CPV), was expressed in tobacco chloroplasts as a fusion protein with cholera toxin B (CTB) and with the green fluorescent protein (GFP). The maximum levels of protein expression were achieved with CTB-2L21, 7.49 mg/g fresh weight (equivalent to 31.1% total soluble protein) and with GFP-2L21, 5.96 mg/g fresh weight (equivalent to 22.6% of total soluble protein, Molina et al. 2004; Fig. 7). The expression levels were also dependent on the age of the plant. Mature plants accumulated the highest amounts of protein when compared to the young and senescent plants. This shows the importance of the plant's harvesting time. Also, the chimeric protein retained the pentamerization and G_{M1}-ganglioside-binding properties of the native CTB. When mice were immunized intraperitoneally with the leaf extracts from CTB-2L21, mice immunized with CTB-2L21 elicited anti-2L21 antibodies able to recognize VP2 protein from CPV (Fig. 8). This is the first report of an animal vaccine and viral antigen expressed in transgenic chloroplasts.

Chloroplast-Derived Therapeutic Proteins

Several human therapeutic proteins that have been expressed in transgenic chloroplasts are listed in Table 2. The expression levels depend on the site of integration, regulatory elements used to enhance transcription/translation efficiency, and the stability of the foreign protein. Genes encoding for proteins as small as 20 amino acids (magainin; DeGray et al. 2001) or as large as 83 kDa

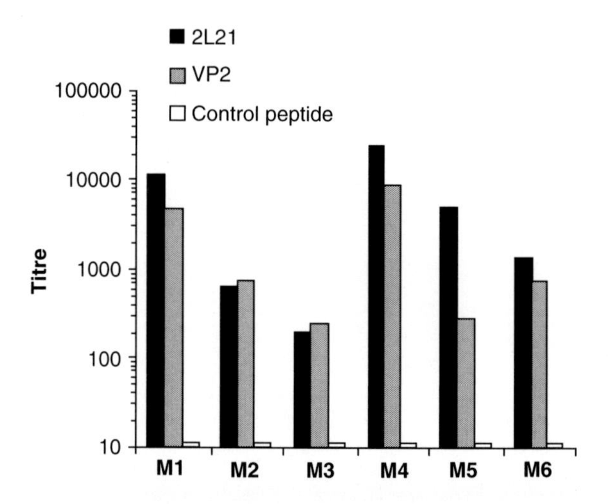

Fig. 8 Titers of antibodies at day 50 induced by plant-derived CTB-2L21 recombinant protein. Balb/c mice were intraperitoneally immunized with leaf extract from CTB-2L21 transgenic plants. Animals were boosted at days 21 and 35. Each mouse received 20 μg of CTB-2L21 recombinant protein. Individual mice sera were titrated against 2L21 synthetic peptide, VP2 protein and control peptide (amino acids 122–135 of hepatitis B virus surface antigen). Titers were expressed as the highest serum dilution to yield twice the absorbance mean of preimmune sera. *M1–M6* mouse 1–6, *2L21* epitope from the VP2 protein of the canine parvovirus, *CTB* cholera toxin B, *VP2* protein of the canine parvovirus that includes the 2L21 epitope

(PA; Daniell et al. 2004b) have been expressed in transgenic chloroplasts. A few illustrations from the Daniell laboratory are described in the following sections.

Human Serum Albumin

Human serum albumin (HSA), the most widely used intravenous protein, is obtained by fractionation of blood serum and accounts for approximately 60% of the total protein in the blood (Fernández-San Millán et al. 2003). The world needs more than 500 tons annually, representing a market value of more than $1.5 billion. HSA was produced in a wide variety of microbial systems but no system is commercially feasible yet. Expression of functional HSA in a transgenic chloroplast expression system is advantageous because of the inexpensive production costs and absence of human pathogens. This protein was expressed in transgenic chloroplasts under the translational control of a Shine Dalgarno sequence, 5′*psbA* region or the *cry2Aa2* UTR. Different expression levels were obtained in transgenic plants with different regulatory sequences; for example, potted transgenic plants regulated by Shine-Dalgarno sequence showed 0.02% TSP, while 7.2% and 0.08% TSP HSA was observed in plants regulated by chloroplast 5′ *psbA* or heterologous *cry2Aa2* UTR, respectively. Expression in seedlings with Shine-Dalgarno control was 0.8% total protein of HSA, while 1.6% TP was observed with 5′ *psbA* control and 5.9% TP under *cry2Aa2* UTR control. The low levels of HSA accumulation in mature plants under the control of SD may be due to excessive proteolytic degradation and poor rates of translation. As 5′ *psbA* region is light-regulated, under continuous light illumination for 50 h, the HSA quantity in mature leaves was maximum: 11.1% of the total soluble protein (Fernández-San Millán et al. 2003). This is 500-fold greater expression than the previous reports of nuclear expression.

HSA was observed to form inclusion bodies in transgenic chloroplasts, thus offering protection from proteolytic degradation (Fig. 9**a–c**). Inclusion bodies can easily be separated from the majority of the cellular proteins by centrifugation, thereby eliminating the need for expensive affinity columns and chromatographic techniques. Properly folded HSA can be recovered from the inclusion bodies after

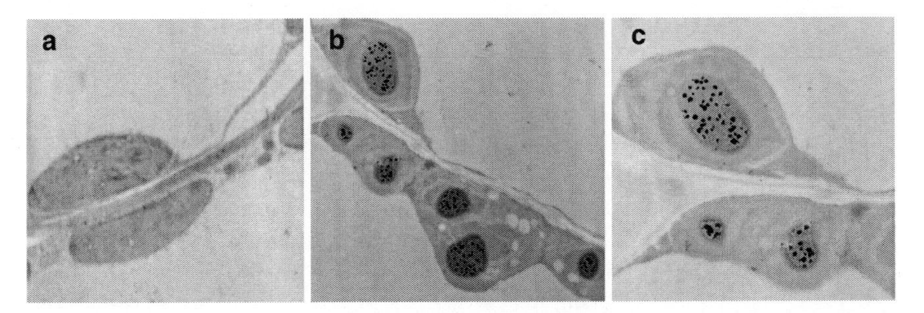

Fig. 9 a–c Study of HSA accumulation in transgenic chloroplasts (**a–c**) Electron micrographs of immunogold labelled tissues from untransformed (**a**) and transformed mature leaves with the chloroplast vector pLDApsbAHSA (**b, c**). Magnifications: A×10,000; B×5,000; C×6,300

denaturation for complete solubilization and in vitro refolding (Fernández-San Millán et al. 2003).

Human Interferon Alpha

Human interferon (IFNα 2b) is used in the treatment of malignant carcinoid tumors and was shown to be very effective in the reduction of tumor size. It has other therapeutic values such as inhibition of viral replication, cell proliferation, and enhancement of immune response, and recently, in treatment of patients suffering from West Nile virus infection. At present, IFNα 2b protein is produced in *E. coli* for commercial use and also requires in vitro processing and purification. The nuclear expression of this protein also resulted in very low expression (0.000017% TSP) in tobacco (Elderbaum et al. 1992). This has altogether made the treatment with interferon very expensive, which on average costs $26,000–$40,000 per year.

Recombinant IFNα 2b was expressed in transgenic chloroplasts using the chloroplast transformation vector integrated with the gene cassette that included IFNα2b gene along with polyhistidine purification tag and a thrombin cleavage site (Fig. 10). Western blots were performed to detect the multimers and monomer of IFNα 2b using interferon alpha monoclonal antibodies and confirm formation of disulfide bonds. Integration of the gene cassette into chloroplast genome was confirmed by Southern blots. ELISA was performed to quantify the IFNα 2b protein and expression up to 18.8% of total soluble protein was obtained (Arlen et al., 2007).

The functionality of IFNα 2b was investigated by its ability to protect HeLa cells against the cytopathic effect of encephalomyocarditis virus (EMC) and through the identification of interferon-induced transcripts (Fig. 11). The chloroplast derived IFNα 2b is found to have the same activity as commercially produced Intron A. The mRNA levels of two genes induced by IFNα 2b (2′-5′ oligoadenylate synthase and *STAT-2*) were tested by RT-PCR using primers specific for each gene. Chloroplast-derived IFNα 2b induced the expression of both genes in a manner similar to commercial IFNα 2b. This confirms that chloroplast derived IFNα 2b is as active as commercially produced Intron A.

Human Insulin-Like Growth Factor-1

Human insulin-like growth factor (IGF-1) has therapeutic value not only in mediating the growth of muscle and other tissues, but its therapeutic value is being currently evaluated in diabetes, IGF-I induced neuroprotection, and in promoting bone healing. The IGF-1 is a naturally occurring single-chain polypeptide with three disulfide bonds, produced in the liver (Torrado and Carrascosa 2003). When produced in *E. coli,* IGF-1 cannot produce the mature form, as disulfide bonds cannot be formed in *E. coli* cytoplasm. Since the IGF-1 has codons suitable for the eukaryotic environment, codon was optimized for chloroplast to increase the levels of expression in transgenic chloroplasts. PCR and Southern blot analysis confirmed

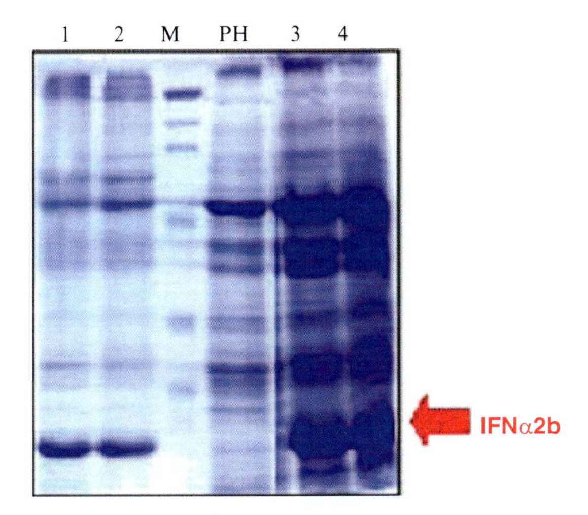

Fig. 10 Coomassie-stained SDS-polyacrylamide gel showing chloroplast transgenic lines expressing IFN α 2b. Lanes 1 and 2: total soluble protein; lanes PH, 3 and 4: total protein

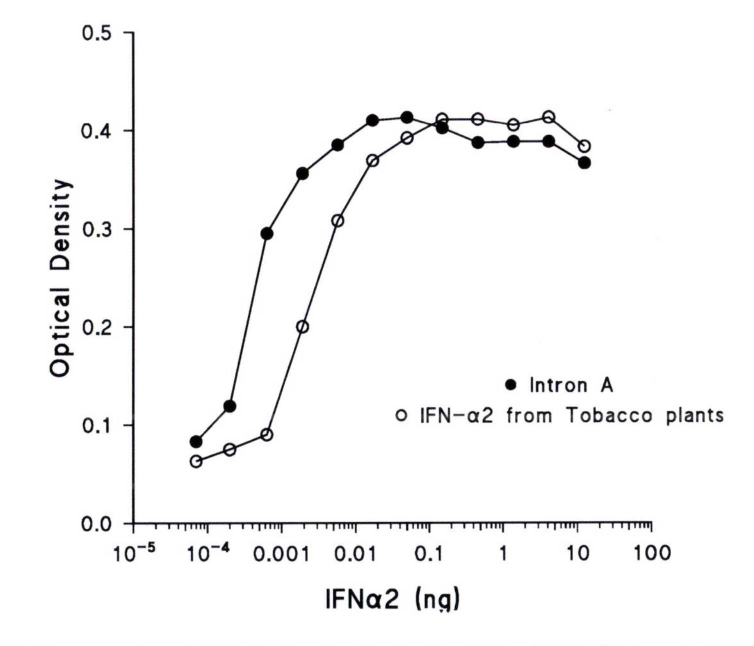

Fig. 11 Demonstration of IFNα 2b functionality by the ability of IFNα 2b to protect HeLa cells against the cytopathic effect of encephalomyocarditis virus. Note that chloroplast derived IFNα 2b is as active as commercially produced Intron A

the chloroplast integration of IGF-1 gene. ELISA was performed to quantify the expression levels of IGF-1 from both native and synthetic genes in transgenic chloroplasts and the expression levels were as high as 32% of the total soluble protein (Ruiz 2002; Fig. 12). However, quantification of expression of IGF-1 was complicated by the zz-tag used for purification and levels of expression should be verified

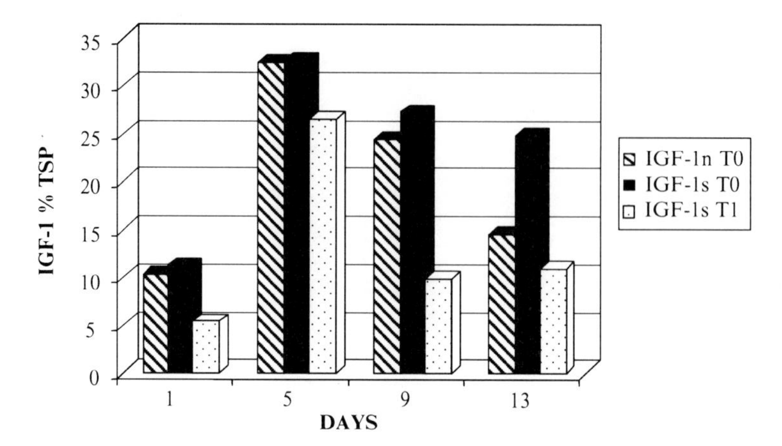

Fig. 12 Expression of IGF-1 in transgenic chloroplasts after continuous illumination for 13 days. IGF-1 expression is shown as a percentage of total soluble protein. IGF-1n is the native gene and IGF-1 s is the chloroplast codon-optimized gene

using antibodies that do not cross-react with the zz-tag. These observations suggest that unlike bacterial translational machinery, chloroplast translation machinery may be quite flexible.

Anti-Microbial Peptide

Anti-microbial peptides (AMPs) are a common component of the innate defense mechanisms in the animal kingdom, which help to combat pathogens and to control normal microbial flora. Magainin, secreted from the skin of the African clawed frog (*Xenopus laevis*), is a broad-spectrum topical agent, a systemic antibiotic and a wound-healing and anticancer agent (Zasloff 1987; Jacob and Zasloff 1994; DeGray et al. 2001). Magainin has affinity for the negatively charged phospholipids in the outer leaflet of the prokaryotic membrane. A magainin analog, MSI-99, was expressed in transgenic chloroplasts and expression levels up to 21.5% of total soluble protein were obtained (DeGray et al. 2001; Daniell et al. 2001a). The effectiveness of the lytic peptide expressed in transgenic chloroplasts was tested with a multidrug resistant Gram-negative bacterium, *Pseudomonas aeruginosa*, and this resulted in 96% growth inhibition of the pathogen (Fig.13). MSI-99 was most effective against *Pseudomonas syringae* requiring only 1 μg/1,000 bacteria based on the study of minimum inhibitory concentration of MSI-99.

Chloroplast-Derived Human Antibody

Monoclonal antibodies for passive immunotherapy have been the most widely studied therapeutic proteins produced in transgenic plants. Though large num-

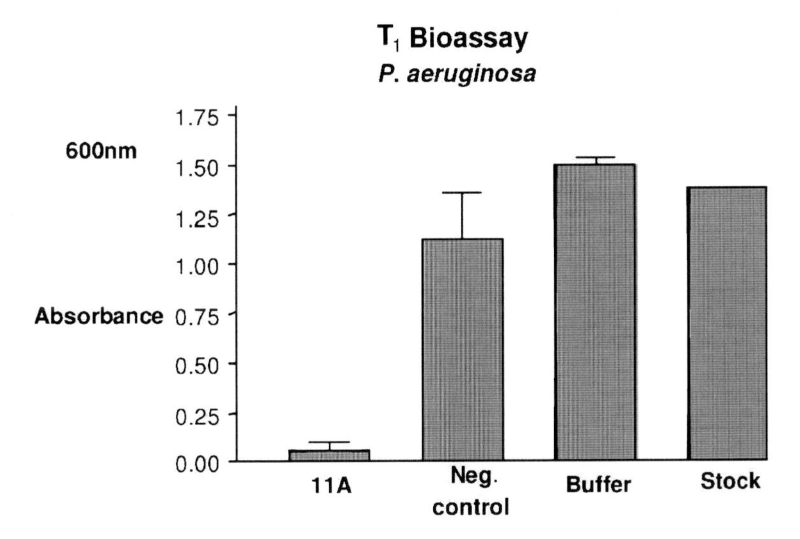

Fig. 13 In vitro bioassay for T1 generation chloroplast transgenic line against *P. aeruginosa*. Bacterial cells from an overnight culture were diluted to A_{600} 0.1–0.3 and incubated for 2 h at 25°C with 100 μ g of total protein extract. Of LB, 1 ml was added to each sample and incubated overnight at 26°C. Absorbance was recorded at 600 nm. Data were analyzed using Graph Pad Prism

bers of therapeutic proteins have been produced in plants, only a few have entered clinical trials. The anti-*Streptococcus mutans* secretory antibody for the prevention of dental caries is the only plant-derived antibody currently in phase II clinical trials (Larrick and Thomas 2001). Based on chloroplast's ability to form fully active and assembled proteins, a codon-optimized and humanized gene encoding a chimeric monoclonal antibody (IgA/G, Guy's 13) under the control of a specific 5′-untranslated region, was used to synthesize monoclonal antibodies in transgenic chloroplasts. Guy's 13 was developed to prevent dental caries, which is caused by *Streptococcus mutans* (Daniell and Wycoff 2001; Daniell et al. 2001a; Daniell 2004). Integration of the chimeric antibody gene into chloroplast genome was confirmed by PCR and Southern blot analysis. Western blot analysis showed the expression of heavy and light chains as well as fully assembled antibody (Fig. 14), suggesting the presence of chaperones for proper protein folding and enzymes for formation of disulfide bond in transgenic chloroplasts. However, expression levels should be enhanced further to facilitate commercialization.

Conclusion

The chloroplast genetic engineering approach is ideal for economical production of vaccine antigens and biopharmaceuticals in an environmentally friendly manner. The proven functionality and the high expression levels of vaccine antigens and therapeutic proteins in transgenic chloroplasts hold the promise for unlimited quan-

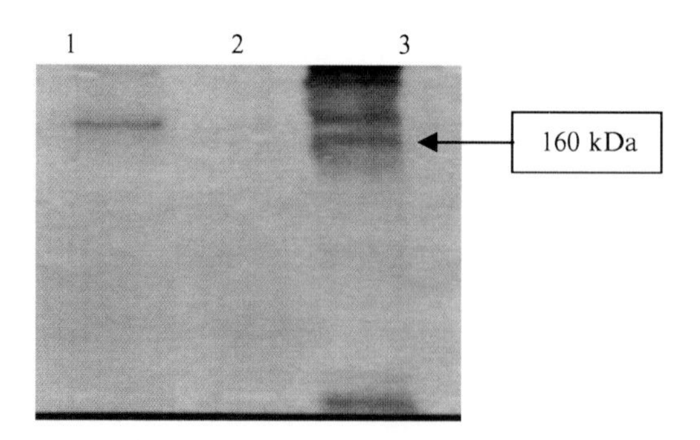

Fig. 14 Western blot analysis of transgenic lines showing the expression of an assembled Guy's 13 monoclonal antibody in transgenic chloroplasts. Lane 1: extract from a chloroplast transgenic line; lane 2: extract from an untransformed plant; lane 3: positive control (human IgA). The gel was run under nonreducing conditions. The antibody was detected with AP-conjugated goat anti-human kappa antibody

tities of production, but the cost of purification is still prohibitive. This can be overcome by the oral delivery of vaccine antigens and therapeutic proteins or by using the novel purification strategies described in this review.

Acknowledgements Results of investigations from the Daniell laboratory are supported in part by the United States Department of Agriculture 3611–21000–017–00D and the National Institutes of Health R01 GM 63879 grants.

References

Arlen PA, Singleton M, Adamovicz JJ, Ding Y, Davoodi-Semiromi A, Daniell H (2008) Effective plague vaccination via oral delivery of plant cells expressing F1-V antigens in chloroplasts. Infect. Immun. 76:3640–3650

Arlen PA, Falconer R, Cherukumilli S, Cole A, Cole AM, Oishi KK, Daniell H (2007) Field production and functional evaluation of chloroplast-derived interferon-alpha2b. Plant Biotechnol. J. 5:511–525

Baillie L (2001) The development of new vaccines against *Bacillus anthracis*. J Appl Microbiol 91:609–613

Bogorad L (2000) Engineering chloroplasts: an alternative site for foreign genes, proteins, reactions and products. Trends Biotechnol 18:257–263

Daniell H (2002) Molecular strategies for gene containment in transgenic crops. Nat Biotechnol 20:581–586

Daniell H (2004) Medical Molecular Pharming. In: Goodman RM (ed) The encyclopedia of plant and crop sciences. Marcel Dekker, New York, pp 705–710

Daniell H, Dhingra A (2002) Multigene engineering: dawn of an exciting new era in biotechnology. Curr Opin Biotechnol 13:136–141

Daniell H, Wycoff K (2001) WO Patent 01–64929, 2001

Daniell H, Guda C, McPherson DT, Xu J, Zhang X, Urry DW (1997) Hyperexpression of an environmentally friendly synthetic polymer gene. Methods Mol Biol 63:359–371

Daniell H, Dhingra A, Fernández-San Millán A (2001a) Chloroplast transgenic approach for production of antibodies, biopharmaceuticals and edible vaccines. In: 12th International Congress on Photosynthesis Vol. S40–04, CSIRO Publishing, Brisbane, Australia, pp 1–6

Daniell H, Lee SB, Panchal T, Wiebe P (2001b) Expression of native cholera toxin B sub-unit gene and assembly as functional oligomers in transgenic tobacco chloroplasts. J Mol Biol 311:1001–1009
Daniell H, Muthukumar B, Lee SB (2001c) Marker free transgenic plants: engineering the chloroplast genome without the use of antibiotic selection. Curr Genet 39:109–116
Daniell H, Streatfield SJ, Wycoff K (2001d) Medical molecular farming: production of antibodies, biopharmaceuticals and edible vaccines in plants. Trends Plant Sci 6:219–226
Daniell H, Wiebe PO, Millan AF (2001e) Antibiotic-free chloroplast genetic engineering: an environmentally friendly approach. Trends Plant Sci 6:237–239
Daniell H (2004) Medical Molecular Pharming. In: Goodman RM (ed) The encyclopedia of plant and crop sciences. Marcel Dekker, New York, pp 705–710
Daniell H, Carmona-Sanchez O, Burns B (2004a) Chloroplast derived antibodies, biopharmaceuticals and edible vaccines. In: Fischer R, Schillberg S (eds) Molecular farming. Wiley-VerlagVCH, Weinheim, Germany, pp 113–133
Daniell H, Watson J, Koya V, Leppla SH (2004b) Expression of *Bacillus anthracis* protective antigen in transgenic chloroplasts of tobacco, a non-food/feed crop. Vaccine 22:4374–4384
De Cosa B, Moar W, Lee SB, Miller M, Daniell H (2001) Over expression of the cry2Aa2 operon in chloroplasts leads to formation of insecticidal crystals. Nat Biotechnol 19:71–74
DeGray G, Rajasekaran K, Smith F, Sanford J, Daniell H (2001) Expression of an antimicrobial peptide via the chloroplast genome to control phytopathogenic bacteria and fungi. Plant Physiol 127:852–862
Dhingra A, Portis AR, Daniell H (2004) Enhanced translation of a chloroplast-expressed RbcS gene restores small subunit levels and photosynthesis in nuclear RbcS antisense plants. Proc Natl Acad Sci U S A 101:6315–6320
Elderbaum O, Stein D, Holland N, Gafni Y, Livneh O, Novick D, Rubinstein M, Sele I (1992) Expression of active human interferon beta in transgenic plants. J Interferon Res 12:449–453
Falconer R (2002) Expression of Interferon alpha 2b in transgenic chloroplasts of a low-nicotine tobacco. M.S. thesis, University of Central Florida
Fernández-San Millán A, Mingo-Castel A, Miller M, Daniell H (2003) A chloroplast transgenic approach to hyper-express and purify Human Serum Albumin, a protein highly susceptible to proteolytic degradation. Plant Biotech J 1:71–79
Figueroa-Soto CG, Guillermo LC, Valenzuela-Soto EM (1999) Immunolocalization of betaine aldehyde dehydrogenase in porcine kidneys. Biochem Biophys Res Commun 258:732–736
Giddings G, Allison G, Brooks D, Carter A (2000) Transgenic plants as factories for biopharmaceuticals. Nat Biotechnol 18:1151–1155
Gomez N, Carrillo C, Salinas J, Parra F, Borca MV, Escribano JM (1998) Expression of immunogenic glycoprotein S polypeptides from transmissible gastroenteritis corona virus in transgenic plants. Virology 249:352–358
Guda C, Lee SB, Daniell H (2000) Stable expression of a protein based polymer in tobacco chloroplasts. Plant Cell Rep 19:257–262
Haq TA, Mason HS, Clements JD, Arntzen CJ (1995) Oral immunization with a recombinant bacterial antigen produced in transgenic plants. Science 268:714–716
Iatham S, Day A (2000) Removal of antibiotic resistance genes from transgenic tobacco plastids. Nat Biotechnol 18:1172–1176
Ivins B, Fellows P, Pitt L, Estep J, Farchaus J, Friedlander A, et al (1995) Experimental anthrax vaccines: efficacy of adjuvants combined with protective antigen against an aerosol *Bacillus anthracis* spore challenge in guinea pigs. Vaccine 13:1779–1783
Jacob L, Zasloff M (1994) Potential therapeutic applications of magainins and other microbial; agents animal origin: antimicrobial peptides. Ciba Found Symp 186:197–223
Joellenback LM, Zwanziger LL, Durch JS, Strom BL (eds) (2002) “Anthrax vaccine manufacture” in the anthrax vaccine. Is it safe? Does it work? National Academy, Washington, DC, pp 180–197
Kaufmann AF, Meltzer MI, Schmid GP (1997) The economic impact of a bioterrorist attack: are prevention and post attack intervention programs justifiable? Emerg Infect Dis 3:83–94
Klaus SM, Huang FC, Golds TJ, Koop HU (2004) Generation of marker-free plastid transformants using a transiently cointegrated selection gene. Nat Biotechnol 22:225

Kumar S, Dhingra A, Daniell H (2004) Plastid-expressed betaine aldehyde dehydrogenase gene in carrot cultured cells, roots, and leaves confers enhanced salt tolerance. Plant Physiol. 136: 2843–2854

Langeveld JP, Casal JI, Osterhaus AD, Cortes E, de Swart R, Vela C, Dalsgaard K, Puijk WC, Schaaper WM, Meloen RH (1994) First peptide vaccine providing protection against viral infection in the target animal: studies of canine parvovirus in dogs. J Virol 68:4506–4513

Langeveld JP, Kamstrup S, Uttenthal A, Strandbygaard B, Vela C, Dalsgaard K, Beekman NJ, Meloen RH, Casal JI (1995) Full protection in mink against enteritis virus with new generation canine parvovirus vaccines based on synthetic peptide or recombinant protein. Vaccine 13:1033–1037

Larrick JW, Thomas DW (2001) Producing protein in transgenic plants and animals. Curr Opin Biotechnol 12:411–418

Lee SB, Kwon HB, Kwon SJ, Park SC, Jeong MJ, Han SE, Byun MO, Daniell H (2003) Accumulation of trehalose within transgenic chloroplasts confers drought tolerance. Mol Breed 11:1–13

Leelavathi S, Reddy VS (2003) Chloroplast expression of His-tagged GUS-fusions: a general strategy to overproduce and purify foreign proteins using transplastomic plants as bioreactors. Mol Breed 11:49–58

Mason HS, Lam D, Arntzen CJ (1992) Expression of hepatitis B surface antigen in transgenic plants. Proc Natl Acad Sci U S A 89:11745–11749

Mason HS, Haq TA, Clements JD, Arntzen CJ (1998) Edible vaccine protects mice against *Escherichia coli* heat-labile enterotoxin (LT): potatoes expressing a synthetic LT-B gene. Vaccine 16:1336–1343

Molina A, Hervas-Stubbs S, Daniell H, Mingo-Castel AM, Veramendi J (2004) High yield expression of a viral peptide animal vaccine in transgenic tobacco chloroplasts. Plant Biotechnol 2:141–153

Ruiz G (2002) Optimization of codon composition and regulatory elements for expression of the human IGF-1 in transgenic chloroplasts. MS thesis, University of Central Florida

Singleton ML (2003) Expression of CaF1 and LcrV as a fusion protein for a vaccine against *Yersinia pestis* via chloroplast genetic engineering. MS thesis, University of Central Florida

Tackaberry E, Dudani A, Prior F, Tocchi M, Sardana R, Altosaar I, Ganz PR (1999) Development of biopharmaceuticals in plant expression systems: cloning, expression and immunological reactivity of human cytomegalovirus glycoprotein B (UL55) in seeds of transgenic tobacco. Vaccine 17:3020–3029

Thanavala Y, Yang Y, Lyon P, Mason HS, Arntzen C (1995) Immunogenicity of transgenic plant-derived hepatitis B surface antigen. Proc Natl Acad Sci U S A 92:3358–3361

Titball RW, Williamson ED (2001) Vaccination against bubonic and pneumonic plague. Vaccine 19:4175–4184

Torrado J, Carrascosa C (2003) Pharmacological characteristics of parenteral IGF-I administration. Curr Pharm Biotechnol 4:123–140

Urry DW, McPherson DT, Xu J, Daniell H, Guda C, Gowda DC, Jing N, Parker TM (1996) Protein based polymeric materials: synthesis and properties. In: Salamone JC (ed) The polymeric materials encyclopedia: synthesis, properties and applications. CRC Press, Boca Raton, FL, pp 2645–2699

Vivek BS, Ngo QA, Simon PW (1999) Evidence for maternal inheritance of the chloroplast genome in cultivated carrot (*Daucus carota* L. ssp. sativus). Theor Appl Genet 98:669–672

Williamson ED, Eley SM, Stagg AJ, Green M, Russell P, Titball RW (1997) A sub-unit vaccine elicits IgG in serum, spleen cell cultures and bronchial washings and protects immunized animals against plague. Vaccine 15:1079–1084

Zasloff M (1987) Magainins, a class of antimicrobial peptides from *Xenopus* skin: isolation, characterization of two active forms, and partial cDNA sequence of a precursor. Proc Natl Acad Sci U S A 84:5449–5953

Zhang Q, Liu Y, Sodmergen (2003) Examination of the cytoplasmic DNA in male reproductive cells to determine the potential for cytoplasmic inheritance in 295 angiosperm species. Plant Cell Physiol 44:941–951

Production of Antibodies in Plants: Approaches and Perspectives

K. Ko, R. Brodzik, and Z. Steplewski

Contents

Abstract Advances in molecular biology, immunology, and plant biotechnology have changed the paradigm of plant as a food source to so-called plant bioreactor to produce valuable recombinant proteins. These include therapeutic or diagnostic monoclonal antibodies, vaccines, and other biopharmaceutical proteins. The plant as a bioreactor for the production of therapeutic proteins has several advantages, which include the lack of animal pathogenic contaminants, low cost of production, and ease of agricultural scale-up compared to other currently available systems. Thus, plants are considered to be a potential alternative to compete with other systems such as bacteria, yeast, or insect and mammalian cell culture. Plant production systems, particularly therapeutic antibodies, are very attractive to pharmaceutical companies to produce the antibodies in demand. Currently, we have successfully

K. Ko(✉)
Department of Biological Science, College of Natural Sciences, Wonkwang University, Iksan, Jeonbuk 570-749, Korea
e-mail: ksko@wku.ac.kr

R. Brodzik and Z. Steplewski
Biotechnology Foundation Labs, Thomas Jefferson University, Philadelphia, PA 19107, USA

A.V. Karasev (ed.) *Plant-produced Microbial Vaccines.*
Current Topics in Microbiology and Immunology 332

developed a plant system for production of anti-rabies monoclonal antibody and anti-colorectal cancer monoclonal antibody. The effective plant production system for recombinant antibodies requires the appropriate plant expression machinery with optimal combination of transgene expression regulatory elements, control of posttranslational protein processing, and efficient purification methods for product recovery. However, there are several limitations that have to be resolved to establish the efficient plant system for antibody production. Here, we discuss the approaches and perspectives in plant systems to produce monoclonal antibody.

Introduction

Over the 30 years since Kohler and Milstein (1975) reported production of monoclonal antibody (mAb) using hybridoma cell line, many mAbs have been produced in diverse expression systems (Chadd and Chamow 2001; Ma et al. 2003; Ko and Koprowski 2005). Antibodies have been developed mainly for tumor therapy (Harris 2004), but it is expanding to other diseases such as pandemic infectious diseases and bioterrorism agents (Casadevall et al. 2004). Antibody itself has a number of advantages such as low toxicity and high specificity, and it can be structurally modified to different forms to be feasible for diverse approaches. The antibodies can be directly armed with radionuclides, toxins, or cytokine for control of tumor or infected cells. In 2001, the antibody market value was US $2 billion. The monoclonal antibody market was one of the fastest growing therapeutic protein areas between 2003 and 2004 (Gomord et al. 2004). The antibody market value was nearly $5 billion in 2005. In 2010, the market will be expanded to more than $30 billion. Antibodies are produced in hybridoma systems or other mammalian culture systems. However, the limited capacity of these systems and their high production cost hamper the attempts to meet such increasing demand of antibodies. Thus, many companies are interested in alternative economically feasible production approaches. Advanced immunology and bioengineering led us to produce monoclonal antibody in variable expression systems, such as *Escherichia coli*, yeast, insect, and mammal cells (Chadd and Chamow 2001). Since expression of antibody in transgenic plants was first described by (Duering 1988; Hiatt et al. 1989), different antibodies and their derivatives have been expressed in plant systems (Fig. 1). The use of transgenic plants for the production of monoclonal antibody has many potential economic and safety advantages (Ma et al. 2003;

Fig. 1 (continued) Mayfield et al. 2003). The different forms of antibodies can be chosen to express in plant on intended antibody-based applications such as therapy and diagnosis. When ideal properties for applications is mainly high affinity for the targeted antigen such as certain molecules to detect, pathogens to inhibit infection, or toxins to neutralize by physically blocking ligand-receptor interactions, antibody derivatives without Fc regions are desirable. In other words, when the antibody-dependent cellular cytotoxicity (ADCC) where the Fc region of the antibody is essential, a full-size or large Fc region is chosen

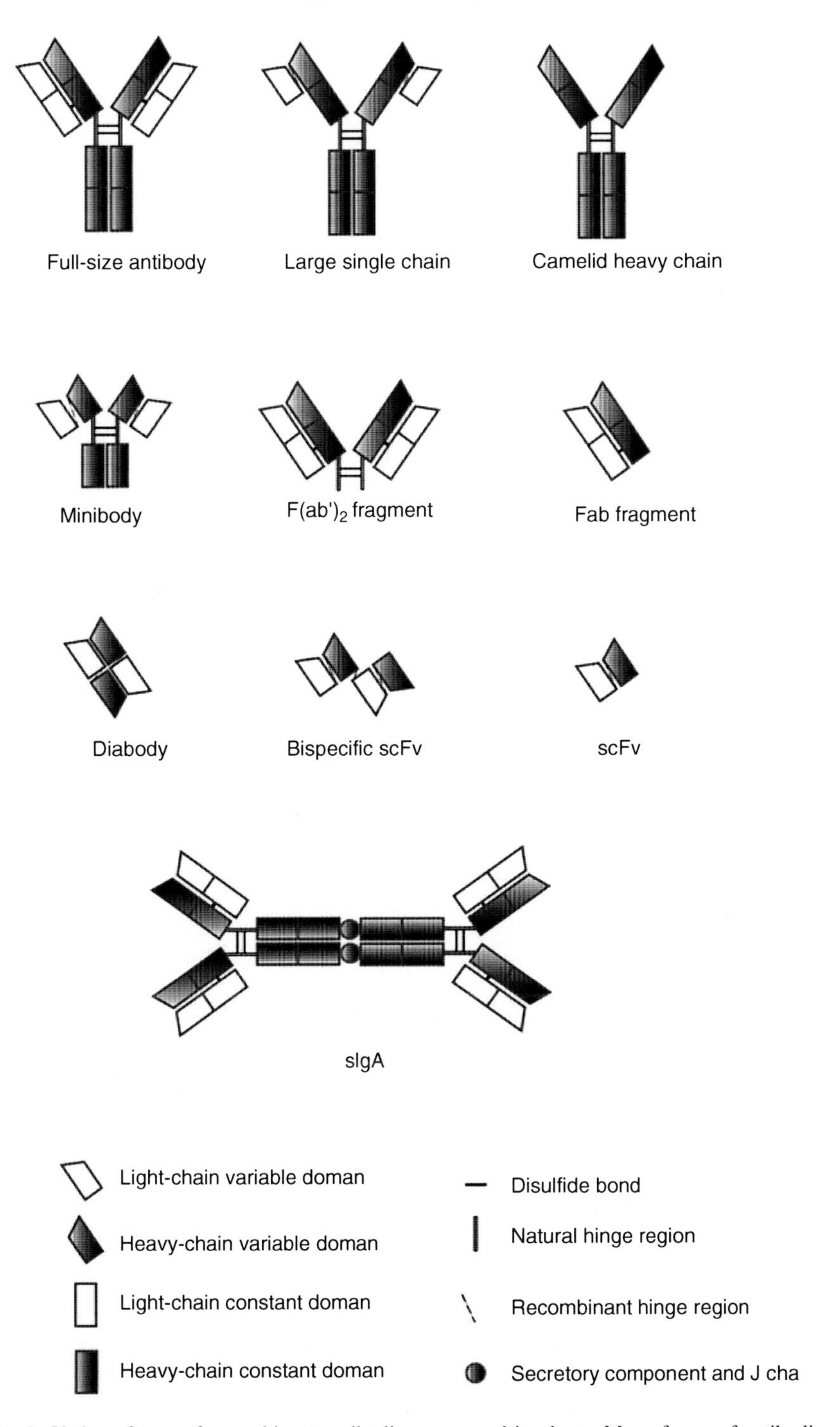

Fig. 1 Various forms of recombinant antibodies expressed in plants. Many forms of antibodies have been expressed in plant systems, including full-size antibody (Hiatt et al. 1989) and its derivatives (Conrad and Fiedler 1998; Fischer et al. 1999; Peeters et al. 2001; Jobling et al. 2003;

Gomord et al. 2004; Ko and Koprowski 2005), including large scale-up, the ease to manipulation, and lack of human pathogenic contaminants. There is no doubt about the capability of the plant expression system for production of antibody since plants are advanced eukaryotic organisms that are able to perform posttranslational modifications. Plant cells correctly assemble and fold antibodies with disulfide bridges and *N*-glycosylation similar to the parental antibody produced in mammalian cell. Although the plant system has economic and safety advantages over other systems, there are several obstacles such as plant-specific *N*-glycosylation, purification costs, environmental impact, and public acceptance of plant-made therapeutic proteins. To obtain full advantage of the plant expression system, it is essential to understand the limitations of current approaches to using plant systems for antibodies and developing novel technology to overcome the remaining hurdles. This chapter discusses current approaches and perspectives on plant systems to produce antibodies to resolve these problems and shows the potential for the use of plants as bioreactors.

Antibody and Its Therapeutic Activity

IgG antibodies are large glycoprotein molecules composed of four polypeptides: two heavy chains and two light chains assembled by disulfide bonds and attached with *N*-glycans (Fig. 2). The heavy and light chains are composed of the variable and constant regions. The amino-terminal domain of light and heavy chains is variable in sequence and therefore termed V_L and V_H, whereas the domain with the constant sequence of each chain is termed C_L and C_H. The variable and constant regions of antibody have two distinctive functions: one is to bind foreign agents

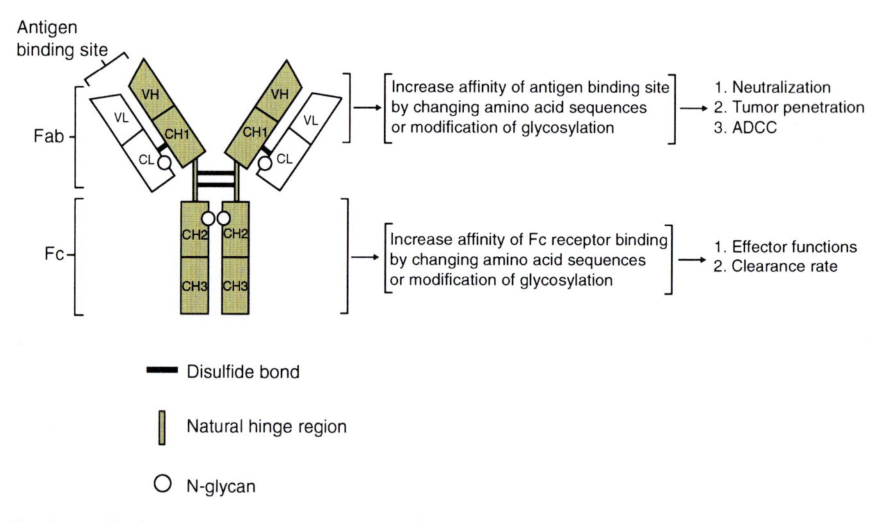

Fig. 2 Antibody structure and its tailor-made functional components

specifically and the other is to recruit various cells and molecules to destroy the foreign agents and pathogens as the antibody binds them. These biological activities of the antibodies allow a wide range of potential applications for immunotherapy and diagnosis.

Antibody-Based Therapies for Cancer

Antibodies have Fab and Fc regions with a specific ability to recognize tumor antigens highly expressed by tumor cells and to recruit immunological effector cells such as natural killer cells and macrophages to disrupt tumor cells, respectively (Herlyn et al. 1980; Houghton and Scheinberg 2000). Thus, antibody therapy is considered a major protein therapeutic for cancer. The mechanisms of this destruction are mainly antibody-dependent cellular cytotoxicity (ADCC), by which immune cells are recruited to kill target tumor cells and complement-dependent cytotoxicity (CDC) (Fig. 3). In some cases, mAbs bound to tumor cells may generate transmembrane signals that directly alter or control tumor growth, potentially leading to growth arrest and apoptosis (Vietta and Uhr 1994; Vuist et al. 1994). Unlabeled antibodies with no attachment of any drug or radioactive material show significant efficacy in treatment of breast cancer, colorectal cancer, non-Hodgkin's

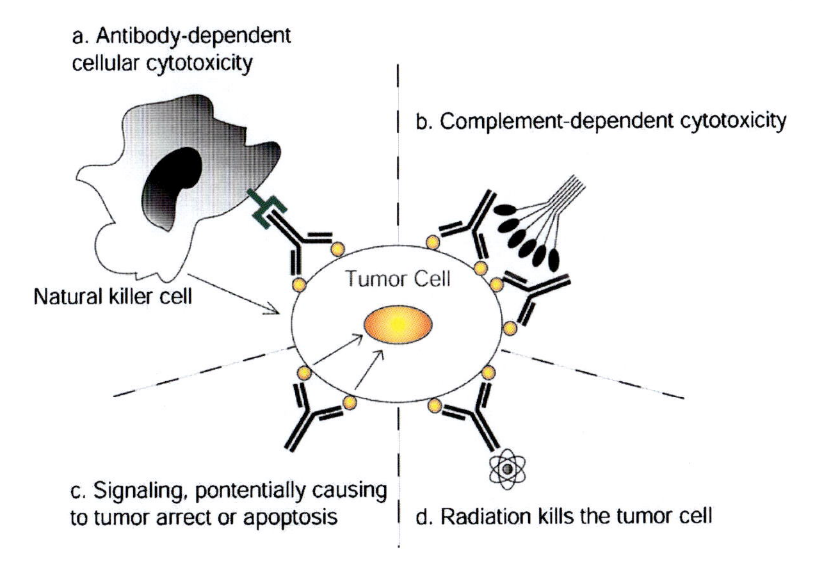

Fig. 3 a-b Mechanisms of anti-tumor activity by antibodies. Monoclonal antibodies recognize antigens on the tumor cells. **a**: Monoclonal antibody binds to Fc receptors on the effector cells, recruiting the effector cells to destroy tumor cells. **b**: Monoclonal antibody bound to antigen activates complement deposition, leading to the lysis of tumor cells. **c**: Monoclonal antibody bound to tumor cells generates transmembrane signals directly altering or controlling tumor growth. **d**: Monoclonal antibody can be armed with a radioisotope, other chemical drugs, or toxins to deliver their cytotoxic agents

Table 1 Antibodies with approval for treatment of cancer. (Modified from Carter 2001; von Mehren et al. 2003; Harris 2004; Gomord et al. 2004)

Antibody name/type	Tumor types	Antigen category/name	Year approved
Edrecolomab/muIgG2a	Dukes C CRC	Growth factor receptors/ EpCAM (GA733–2)	1995
Rituximab/chIgG$_1$	Lymphocytic leukemia	Hematopoietic/CD20	1997
Trastuzumab/huIgG$_1$	Breast cancer	Growth factor receptors/ HER2/Neu	1998
Gemtuzumab ozo-gamicin/humIgG$_4$	Acute myelocytic leukemia (AML)	Hematopoietic/CD33	2000
Alemtuzumab/huIgG$_1$	Chronic lymphocytic leukemia (CLL)	Hematopoietic/CD52	2001
Ibritumomab tiuxetan/ muIgG$_1$	Non-Hodgkin lymphoma	Hematopoietic/CD20	2002
Tositumomab/muIgG2a	Non-Hodgkin lymphoma	Hematopoietic/CD20	2003
Cetuximab/chIgG$_1$	Colorectal cancer	Growth factor receptors/ HER1	2004
Bevacizumab/humIgG$_1$	Colorectal cancer	Angiogenesis and stromal antigen/VEGF	2004

Table 2 Human infectious diseases which antibodies have been developed to control. (From Ma et al. 1995; Ko Ko et al. 2003; Mett et al. 2005)

Disease	Microorganism	Target
Antrax*	*Bacillus anthracis*	Protective antigen
Botulism	*Clostridium Botulinum*	Botulinum neurotoxins
Ebola virus	Ebolar virus	Ebola glycoprotein
Rabies*	Rabies virus	Virus glycoproteins
RSV infection	Respiratory syncytial virus	F glycoproteins
Smallpox	Variola major	14-kDa protein encoded by the A27 gene
West Nile virus	West Nile virus	Viral envelope (E) protein
Dental caries*	*Streptococcus mutans*	Streptococcal antigen I/II

*Monoclonal antibodies for these diseases have been expressed in transgenic plants (Ma et al. 1995; Ko et al. 2003; Mett et al. 2005)

lymphoma, and chronic lymphocytic leukemia through these anti-cancer mechanisms (Table 1). Monoclonal antibodies conjugated to radioactive materials, toxins, and chemotherapy drugs also show efficacy in non-Hodgkin's lymphoma and acute myeloid leukemia (Fig. 3). Thus, antibodies are considered smart guided missiles. Several anti-cancer mAbs have been produced in plant systems (Verch et al. 1998; Vaquero et al. 2002; Ko et al. 2005).

Passive Immunization for Infectious Diseases

Passive antibody therapies for infectious diseases caused by the viral, bacterial, fungal, and parasitic microbes are currently gaining interest in research and clinical fields (Casadevall et al. 2004) (Table 2). Although many vaccines against infectious diseases have been developed and used, the limited vaccine production and immunization against currently emerging infectious diseases hinders protection of unexpectedly infected patients. The increase in drug-resistant microorganisms is driving the development of antibody-based immunotherapeutic applications for the prevention and treatment of infectious diseases.

Advanced cloning of human antibodies from combinatorial libraries can rapidly select mAbs carrying intended affinity against infectious diseases (Burton and Barbas 1994; Winter et al. 1994), giving advantages over vaccines and antimicrobial drugs. The advantage of antibody-based therapy is the high specificity and thus low toxicity compared to conventional drugs. In contrast to anti-cancer immunotherapy, which relies on identifying self-antigens highly expressed in tumor cells, antibody-based passive immunization for infectious diseases benefits from recognizing pathogen-originated antigens, which are largely different from those of the hosts. In infectious diseases caused by a microorganism such as a virus, the high specificity to a certain virus strain does not always provide an advantage since a single highly specific antibody is not satisfactory to control microorganisms with high antigenic variation. The emergence of virus mutants or different virus strain variants encourages the use of low specific antibody activity with binding specificity against broad virus strains. Antibody-based therapies with the multiple low and broad specific antibodies provide a synergistic or additive effect to control viral diseases by variant virus strains (Nosanchuk et al. 2003; Casadevall et al. 2004). In principle, the cocktails of antibodies that are specific for different virus strains is thought to efficiently control virus escaping, as a combination of antiretroviral drugs against HIV-1 infection (Prosniak et al. 2003; Montefiori 2005; Trkola et al. 2005). In the case of bacterial toxin, combining multiple monoclonal antibodies with different targets related to pathogenesis increased the potency of in vivo botulinum toxin neutralization (Nowakowski et al. 2002). Additional advantages of using antibodies are easy modification of the antibody to its derivative structures (Fig. 1). The biological mechanisms of the therapeutic antibody are mainly distinguished by the involvement of the Fc region (Fig. 4). The mechanisms such as toxin or virus neutralization and direct antimicrobial functions have an Fc-independent action, whereas antibody-dependent cellular cytotoxicity have an Fc-dependent action. In the former mechanism, the binding of antibody to a target antigen via the binding region of Fab is sufficient to mediate antimicrobial effects. In contrast, the latter mechanism requires an intact molecule of antibody carrying both Fab and Fc structures. Thus, different forms of antibody derivatives can be chosen for the intended antimicrobial mechanisms (Ko and Koprowski 2005). Neutralizing activity that causes the loss of virus infectivity is essential for the antiviral activity of antibodies. In HIV-1 virus, a combination of three human neutralizing mAb can delay viral rebound after cessation of antiretroviral treatment (Trkola et al. 2005), suggesting that antibodies that neutralize HIV-1 in vitro can suppress the virus in infected individuals. The potency of neutralization is enhanced by increased functional affinity of the mixture antibodies. However, it does not necessarily mean that neutralization is the major antiviral mechanism of protective activity (Fig. 4).

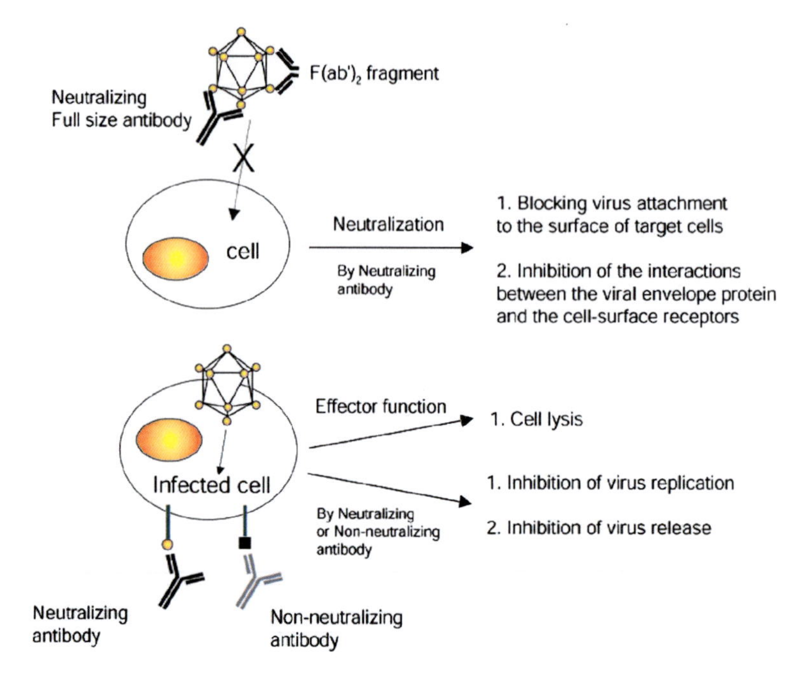

Fig. 4 Mechanisms of antiviral activities of antibodies before and after virus infection. Neutralization is mediated by antibody that binds to surface molecules on viral particle and thus blocking virus entry before virus infection. Antibodies specific for spike proteins of an enveloped virus or other antigens presented by infected cells trigger Fc-mediated effector mechanisms to eliminate infected cells. Neutralizing and non-neutralizing antibodies specific for virus proteins processed in infected cell can give antiviral activity against infected cells, such as cell lysis and inhibiting virus replication and release

Neutralizing IgG1 antibodies with poor effector functions is often ineffective at protection but IgG2a antibodies highly triggering effector functions with the same specificity are effective (Schlesinger et al. 1993), indicating that the mechanisms other than neutralization such as effector functions can be important for protection by neutralizing or non-neutralizing antibodies.

Tailor-Made Antibodies to Improve Efficacy

Antibody-based immunotherapies require particular antibody mechanisms to obtain therapeutic efficacy, as described above. When antibodies are applied to inhibit the activity of certain molecules, pathogens, or toxins by physically blocking ligand–receptor interactions, the ideal properties will be only high affinity of the variable binding region for the targeted antigen. On the other hand, when the desired therapeutic mechanism replies effector functions such as ADCC, complement action, and phagocytosis, both the binding property of mAbs to a target antigen and efficient interactions between the antibody Fc region and Fc receptors are required. In this case, both Fab antigen-binding and Fc constant regions play an equally important role (Fig. 2).

Today, by using advanced molecular biology and immunology techniques, it is possible to redesign the desired properties such as antigen specificity, antigen-binding affinity, and Fc receptor-binding affinity (Fig. 2). To generate mAbs with high antigen specificity and antigen-binding affinity, the mAb antigen-binding site is altered by site-directed mutagenesis (Schier et al. 1996) or random mutagenesis with yeast surface display or phage display (Colby et al. 2004; Brockmann et al. 2005) or glycosylation on variable region of the mAb is modified (Tachibana et al. 1992; Coloma et al. 1999). Glycosylation of Asn^{58} of the V_H of anti-dextran mAb increased the affinity of the antibody for antigen approximately tenfold while carbohydrate at Asn^{60} of the V_H only increased the affinity threefold (Wright et al. 1991), indicating that the glycosylation position affects affinity of variable antibody regions. The structure of the V region carbohydrate can be associated with differences in binding specificity and affinity (Matusuuchi et al. 1981; Tachibana et al. 1992). The enhanced affinity improves antibody neutralizing activity, which is important for inhibition of virus infection (Burton 2002). Thus, simple modification of the glycosylation position might improve the antibody's neutralizing activity. In tumor immunotherapy, the high affinity for tumor antigen enhances cytotoxicity of bispecific antibody against both Fcγ RIII and cancer antigen (McCall et al. 2001). The high affinity bispecific antibodies are retained longer by tumor cells, thus allowing more time for the leukocytes to bind to the available anti-Fcγ RIII binding domain of the bispecific antibody, enhancing its cytotoxicity. However, high affinity may not always be helpful when rapid and complete tumor penetration is essential for immunotoxins since the higher the affinity with the higher antigen density, the greater the binding site barrier, which severely hinders the diffusion of antibodies throughout the tumor mass (Juwied et al. 1992; Weiner and Carter 2005). In this case, relatively low-affinity IgG that relies on a high surface density of antigen on tumor cells can be tuned up to enhance both targeting selectivity and antibody diffusion (Carter 2001).

When the therapy requires the expanded presence of antibody in the blood stream, a slow blood clearance rate is required. The antibody's blood clearance rate can be optimized to enhance antibody efficacy by modification of sequences or glycosylation of mAb Fc regions that affect antibody structural confirmation (Vaccaro et al. 2005; Weiner and Carter 2005; Fig. 2). On the other hand, in immunotoxin-based therapy, rapid clearance of immunotoxins is preferable for reduction of the side effect caused by sustained existence of circulating immunotoxins in body. In glycosylation, lack of galactose decreases the half-life through binding of agalactosyl IgG to mannose-binding protein or mannose receptors (Malhotra et al. 1995; Wright and Morrison 1998). In plants, the half-life of high mannose-type mAb with the heavy chain fused to KDEL, the endoplasmic reticulum (ER) retention motif retaining mAb in ER has rapid blood clearance, indicating that modification of glycosylation in plants might be a useful strategy to regulate the blood clearance rate (Ko et al. 2003).

Also, the altered glycosylation pattern on Fc regions of mAb increases ADCC by enhancing the interaction between Fc regions and Fc receptors (Shields et al. 2002). For example, Umana et al. (1999) showed that high levels of bisected, nonfucosylated glycan in the Fc region enhanced antibody-dependent cellular cytotoxicity. When major effector functions of a given mAb are mainly mediated by ADCC, the lack of fucosylation might increase the mAb anti-tumor activity by enhancing ADCC activity (Herlyn et al. 1986; Shields et al. 2002). In anti-colorectal cancer mAb CO17–1A, mAb-mediated tumor inhibition is due mainly to ADCC activity (Herlyn et al. 1980). Thus, the ADCC increased by using an altered glycosylation pattern on the Fc region of mAb is beneficial for mAb CO17-1A. The use of a plant system offers scalability and shorter timelines to produce antibody. In addition, simple manipulation of plant glycosylation machinery through subcellular localization of antibody and blocking/insertion enzymes related to glycosylation can refine the plant system to enhance the tailor-made quality of antibody. This issue will be discussed later. To maximize the advantages of the plant system, before the transgenes of antibodies are integrated into plant genome, tuning up the intended biological characteristics of antibody should be considered by designing an antibody structure and modifying the sequences and glycan composition.

Current Approach to Express Antibodies in Plants

In plants, to achieve the expression of antibodies, the cDNAs of heavy and light chains of monoclonal antibody from a hybridoma cell are cloned and transferred into the plant genome. Two general methodological categories to transfer the cDNAs to be expressed in plant cells are stable and transient expression. For stable expression, *Agrobacterium*-mediated transformation or particle bombardment are mainly applied as a representative method to stably insert DNA fragments encoding both heavy and light chains to the nuclear or other compartment genomes in plants. The heavy and light chain genes can be introduced to plant cells at the same time by using a single transformation event with one plant expression binary vector (Ko et al. 2003). Often, the genes are separately inserted and expressed in plant lines and co-expressed by crossing these individual lines (Ma et al. 1995; Fig. 5a). Transgenic lines that contain nucleus carrying both inserted genes are further self-crossed to obtain homozygous transgenic plants expressing monoclonal antibody. The advantage of this approach is that lines with high expression of heavy or light chains can be screened and crossed to generate transgenic lines expressing both heavy and light chains with other antibody elements for secretory IgA (Ma et al. 1995; Fig. 5c). This conventional crossing approach can be applied to express multiple monoclonal antibodies and modify glycosylation (Fig. 5b and d, respectively).

The gene can be inserted into chloroplast genome to generate so-called chloroplast transgenic plants in which chloroplasts express and properly fold functional antibodies with disulfide bonds (Daniell 2002a). The advantages of chloroplast genome-based transgene expression are the no-position effect, no gene silencing, and high expression and accumulation. Although expression of the heterologous transgenes is high in chloroplast, it might not be suitable to express monoclonal antibody, which requires glycosylation for the full biological activity since chloroplasts lack glycosylation machinery in nature. Thus, the possibility of establishing glycosylation pathways in the chloroplast compartment has been investigated by incorporating the multiple genes for glycosylation processing (H. Daniell, personal communication). If the equipping glycosylation machinery in the chloroplast com-

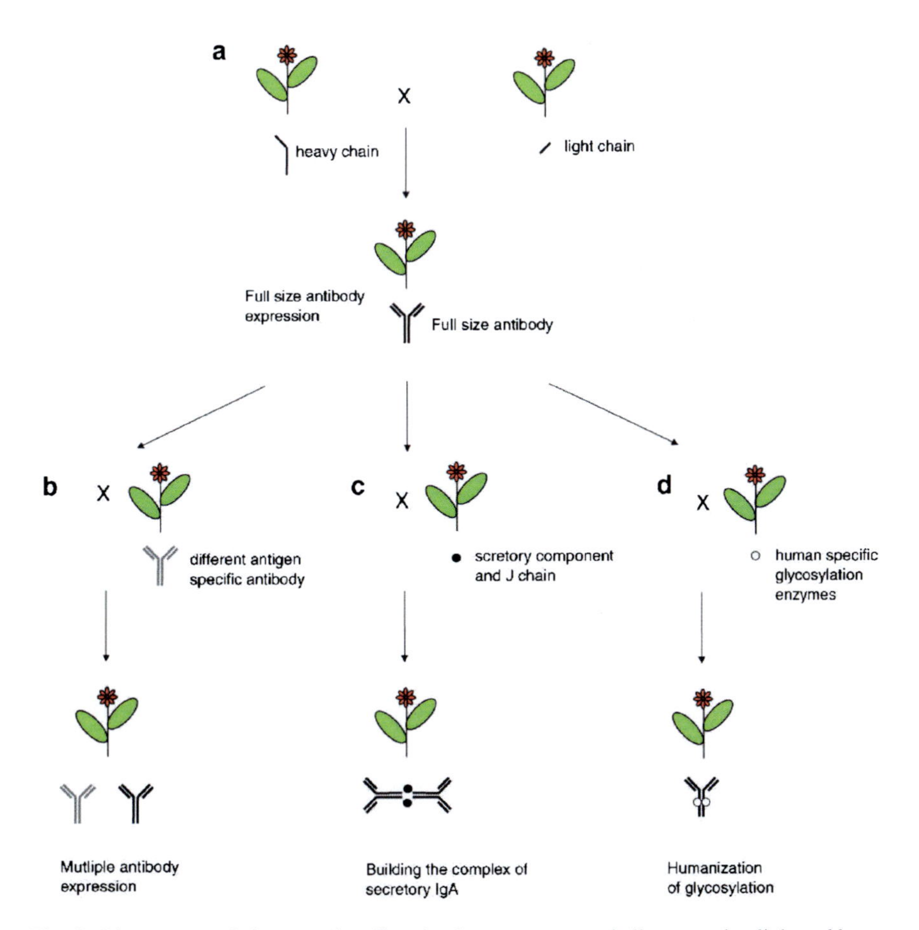

Fig. 5 Diverse uses of plant crossing. Crossing between transgenic lines carrying light and heavy chains, respectively, can generate a transgenic line expressing full-size antibody. This crossing strategy can be efficiently applied to express multiple monoclonal antibodies and a highly complex form of antibody such as a secretory IgA. Glycosylation of monoclonal antibody can be modified by co-expression of antibody and a human-specific glycosylation enzyme such as a human galactosyltransferase in a single transgenic line obtained from crossing lines expressing antibody and the human enzyme (Bakker et al. 2001). This easy manipulation by crossing is another advantage over the other production systems

partment is possible, the chloroplast transgenic plants can become ideal choices for production of most monoclonal antibodies.

For transient expression, agroinfiltration (Vaquero et al. 2002) and recombinant plant viruses (Verch et al. 1998) are mainly used to produce antibodies without generating transgenic plants. Agroinfiltration is generally used as an evaluation method to determine the efficiency of expression cassettes and the activity of recombinant proteins over a short period time before generating transgenic plants. This agroinfiltration method can be used successfully for large-scale production systems. Full-size monoclonal antibodies have been efficiently produced by agroinfiltration systems (Kathuria et al. 2002; Vaquero et al. 2002; Hull et al. 2005).

Plant viruses have been used as protein-expression vectors carrying transgenes both for vaccine delivery and therapeutic protein production (Koprowski and Yusibov 2001). The advantage of this viral vector system is high level of expression, since the virus infection is rapid and systemic to whole plant tissue after inoculation, and easy purification of virus particles. As a production system, the antibodies have been expressed in *N. benthamiana* using the Tobacco Mosaic Virus (TMV) vector (Verch et al. 1998; McCormick et al. 1999). For full-size monoclonal antibody expression, two TMV vectors carrying the heavy and light chain of CO17–1A monoclonal antibody and anti-colorectal cancer antibody were co-infected to *N. benthamiana* (Verch et al. 1998). The expressed light and heavy chains were correctly assembled to full-size monoclonal antibody. For single-chain antibody (scFv), this virus system has a rapid reproduction of customized scFv for cancer patients who usually have unique epitopes to be targeted. This approach could provide personalized immunotherapies for diseases such as non-Hodgkin's lymphoma using easy virus vector-producing scFv and transient expression in a short period of time (McCormick et al. 1999).

This viral vector system requires virus transcript inoculation on each plant tissue since systemic movement of viruses is often limited depending on the size of the integrated transgenes. Also, high mutation and deletion rates are often obtained on the foreign gene during plant RNA virus replication (Smith et al. 1997). The main cause of low systemic infectivity is often the large size fused to coat protein, which hampers efficient viral particle cell-to-cell movement. Marillonnnet et al. (2005) recently developed systemic *Agrobacterium tumefaciens*-mediated transfection of viral replicons for efficient transient expression in plants. *Agrobacterium*-mediated delivery of RNA viral vectors as DNA precursors results in simultaneous gene amplification in all leaves of a plant. This method relies on *Agrobacterium*-infiltration delivering viral vectors. Thus, high expression of heterologous genes in individual bacterial infected cells and consequently no requirement of systemic viral particle movement resolve the gene size-limitation problem and provide synchronous and faster expression. This transient expression process yields as much as 4 g of recombinant protein (green fluorescent protein) per kilogram of fresh leaf biomass in *N. benthamiana* and up to 2.5 g/kg of tobacco (*N. tabacum*), can be applied to other plant species (Gleba et al. 2005; Marillonnnet et al. 2005).

Other Applications of Antibodies Expressed in Plants

Antibody is biologically active protein with its specific binding properties to target antigens resulting in physically or biologically blocking the activities of the antigens. These unique biological activities are applied to other applications such as the sensitive detection and removal of environmental contaminants and industrial purification and processes. For agricultural research, the applications include engineering antibody-mediated resistance to plant disease and immunomodulation of physiological processes (Stoger et al. 2002). In this section, we discuss potential approaches for agricultural applications rather than medical applications using antibody expressed in transgenic plants.

Disease Resistance in Plants

Plant pathogens are a great and growing threat to crop production worldwide. The current homogenous plant cultivation system is vulnerable to the outbreak of epidemics. Conventional plant breeding using sexual crossing of crops is often limited with a narrow choice of resistance genes in plants that are transferred to their elite line. However, plant genetic engineering can generate transgenic plants resistant to plant pathogens by expression of antimicrobial proteins, pathogen-related proteins, or antisense RNAs against pathogenesis obtained from interspecies. Expression of antibody in plants is another novel approach to obtain disease-resistant plants. Application of antibody to control pathogens in plants relies on its interaction with pathogens and antigens involved in pathogenesis. Expression of several antibodies in plants has been described to reduce infection and symptoms caused by viruses, insects, and animals. This antibody-based disease resistance approach is dependent on the precise localization of antibodies in specific subcellular compartments of plant cells or specific tissues. When scFv against TMV is expressed in the cytosol, it effectively confers resistance despite a lower level of expression than that of secreted antibody (Voss et al. 1995). In contrast to TMV, Mollicutes such as phytoplasma responsible for more than 300 diseases of vegetable, ornamental, and perennial plants are strictly localized in the sieve tubes of the phloem tissue. Transgenic tobacco shoots expressing phytoplasma-specific scFvs secreted to the apoplast grew free of disease symptoms (Le Gall et al. 1998), indicating secreted scFv is more effective than cytosol localized scFv. Baum et al. (1996) proposed that antibody-based resistance can also be applied to obtain plants that are resistant to nematodes. The plant nematode's stylet secretions are essential for the initial steps of pathogenesis. Thus, anti-stylet secretion-specific antibody was expressed and secreted into the intercellular space of plant cells. However, secreted antibody did not effectively reduce nematode infection (Baum et al. 1996). It seems that the enzyme produced from the nematode cannot be inactivated by antibodies at apoplast, proposing that cytosolically localized antibody might provide enhanced resistance since the nematode stylet directly injects pathogenesis-related enzymes into cells at the early stage of infection (Baum et al. 1996). Baum et al. (1996) proposed that in the case of root-knot nematodes, a cellulase from *Meloidogyne incognita* is a potential antigen to be selected as a target for antibody since this enzyme plays an essential role in the migration of the nematode inside the plant (Vrain 1999). Another antibody-mediated approach is the use of recombinant antibodies fused to antimicrobial agents expressed in plants to deliver the antimicrobial proteins to the infection site, where the pathogens are populated at the early stage of infection (Schillberg et al. 2001). An antibody-mediated strategy to obtain disease resistance will be appreciated as an alternative approach to control plant disease free of the many limitations if localization of targeted antibody can be regulated with its high expression level.

Antibody-Mediated Metabolic Engineering in Plants

Antibody-mediated metabolic engineering focuses on development of plant varieties with greater yields of specific products (such as carbohydrates, proteins, and oils), and improved tolerance to environmental stress. Immunomodulation manipulates plant cellular metabolism by antibody-mediated alteration of the protein function in plants. Jobling et al. (2003) described the first application of immunomodulation to efficiently inhibit enzyme activity involved in starch biosynthesis in the potato. In this study, they confirmed that expression of plastid-targeted starch-branching enzyme A neutralizing scFv increases accumulation of high-amylose potato starch in the potato. Antibody-mediated metabolic engineering can be applied to understand changes of molecular structures and protein–protein interactions involving stress tolerance. The conformational dynamism and aggregate state of small heat shock proteins (sHSPs) are essential for the function of this protein on heat stress in plants (Miroshinichenko et al. 2005). The sHSPs are important for thermotolerance of plant cells from the detrimental effects of heat stress. Ectopic expression of scFv antibodies against cytosolic sHSPs was used to generate sHSP loss-of-function mutants, which is the lack of the sHSP assembly in vivo, resulting in a lower survival rate in plant cells under heat-stress conditions. He showed that the ability of sHSPs to assemble into heat-stress granulars (HSGs) as well as the HSG disintegration is a prerequisite of survival of plant under the heat stress conditions. This approach could be applied with direct mutagenesis and homology-dependent gene used to understand important metabolisms in the plant.

Antibody-Mediated Phytoremediation

Phytoremediation is an approach which extracts, sequesters, or detoxifies pollutants in soils and surface waters by using plants (Drake et al. 2002). The most advanced strategy in this field is using metal-absorbing plants to clean up areas polluted with lead, cadmium, and copper. Several plant species, particularly in the *Brassicaceae*, are efficient metal-absorbing plants. Transgenic plants expressing protein detoxifying metal pollutants enhance tolerance of metal pollutants and accumulate greater amounts of these metals, suggesting its potential usage for phytoremediation (Song et al. 2003). The majority of environmental pollutants are pesticides, endocrine-disrupting chemicals that are implicated in reproductive system disorders in exposed animal species. Drake et al. (2002) proposed that the phytoremediation capabilities of plants could be extended to these types of chemicals to neutralize by generating transgenic plants expressing an antibody specific to such chemicals. Two strategies of rhizosecretion-mediated binding and sequestration in leaf tissue could potentially be used in the phytoremediation of any pollutant. It is possible to express a monoclonal antibody neutralizing biologically active pollutants and immune complex formation in situ on the plasma membrane in leaves (Drake et al. 2002).

Plant Species to Produce Monoclonal Antibody

Diverse plant species have been transformed for production of recombinant antibodies (Schiermeyer et al. 2004). The choice of plant species is crucial for successful plant biopharming for antibody production since each plant species has unique characteristics that affect antibody expression, product storage, downstream processes, and the quality of the final antibody products (Table 3). The yield of functional antibodies is the first standard to be considered for the choice of plant species. In tobacco, the main advantages are the high biomass yield and the rapid scale-up by prolific seed production compared to other plant species. In addition, tobacco is a non-food, non-feed plant which can attenuate biosafety concerns. Currently, well established tobacco rhizosecretion systems where antibodies are targeted to the secretory pathway for ease of purification (Borisjuk et al. 1999) or a magnifection system where the TMV agroinoculation delivery vectors enhance the expression level makes this plant species attractive. However, tobacco generates a heterogeneous *N*-glycosylation profile of antibodies reflecting the heterogeneous distribution of antibody in the secretory pathway, which may make it difficult to

Table 3 Plant species used for antibody production

Species	Expression tissues	Storage cost	Other culture system	Glycosylation	Protein expression
Alfalfa	Leaf	High	Difficult	Homologous	High
Algae	Total tissue	High	Easy	Heterologous	Medium
Arabidopsis	Leaf	High	Difficult	Heterologous	Medium
Maize	Seed	Low	Difficult	Heterologous	High
Rape	Seed	Low	Difficult	Heterologous	Low biomass
Rice	Seed	Low	Difficult	Heterologous	Medium
Soybean	Seed	High	Difficult	Heterologous	Low biomass
Tobacco	Leaf	High	Easy	Heterologous	Medium

control standardized antibody quality (Cabanes-Macheteau et al. 1999; Ko et al. 2003, 2005). In contrast, alfalfa produces a relatively homologous *N*-glycan structure with its highly efficient folding and/or secretion machinery and fixes atmospheric nitrogen (D'Aoust et al. 2004), which is an advantage in terms of the cultivation of legumes over other plant species. Although alfalfa contains oxalic acid compounds, which affects downstream processing and produces lower amounts of leaf biomass than tobacco, the high protein level in alfalfa leaf tissues maximizes accumulation of recombinant antibodies per plant biomass.

One of disadvantages of these leafy crops comes from the storage and distribution of the plant leaf product, which is instable unless the leaf tissue is frozen or processed. Antibodies expressed in corn seeds are stable at room temperature for more than 3 years without loss of activity (Stoger et al. 2000). Thus, seeds are advantageous in terms of the cost of grain storage and distribution. However, in contrast to leaf plant products such as tobacco, after harvesting of corn seeds for antibody purification, large volumes of waste tissues such as stems and leaves remain, which might have an impact on the environment and generate extra costs downstream. Although the costs of extraction and purification for corn seeds might be higher than plant leaf materials, a well-established food-processing facility may be able to rapidly start up downstream processing. Several other cereal and legume crops have been used for antibody production. Usage of these food crops for therapeutic antibody production may raise concerns about environmental or biosafety issues because of the potential risk of transgene flow to the nontransgenic food crops and contamination of food products by outcrossing (Ko et al. 2003; Ma et al. 2003; Stoger et al. 2005). As plants for pharmaceutical protein production, non-food and non-feed plants might be the choice to avoid transgene contaminants (Ma et al. 2003). There is no ideal choice of plant species since each of the plants has its advantages and disadvantages. Thus, the choice of host plants should be carefully determined.

Purification of Plant-Derived Antibody

The plant system has many advantages compared with other systems for the biopharming of antibodies. One of main advantages of the plant system is the low cost of production. However, downstream processes such as extraction and purification of proteins, a key step, represents more than half of total production cost. Even if the low cost of upstream production is achieved, the total production cost might not be satisfactory without reducing this downstream cost. Thus, efficient purification procedures for plant production systems are as important as effective expression levels. In a plant expression system, the plant tissue and cells should be disrupted to release antibodies for purification since antibody is expressed and localized within the cells. In addition, antibodies should be recovered with a removal of cell debris, noxious chemicals, and contaminants. Currently, for purification of antibody expressed in plants, an affinity purification protocol exploiting protein A-based matrices is mainly used. Large-scale purification has been using protein A streamline chromatography (Valdes et al. 2003). However, the cost of the protein A is prohibitive and fine plant cell debris is difficult to remove from the plant leaf extracts before applying to protein A affinity chromatography, often blocking the column process, consequently reducing durability of protein A column. Although several affinity tagging systems such as the histine tag and intein fusion expression have been applied for purification of recombinant proteins, these systems do not resolve the problems caused by plant cell debris. Therefore, oleaginous plants such as a rapeseed oil are useful hosts for monoclonal antibody production and purification since the oil bodies can be applied to simplify the first steps of antibody isolation (Boothe et al. 1997; Seon et al. 2002). Transgenic plant expresses both protein A fused to oleosin anchored into oilbodies and monoclonal antibody in two distinct cellular compartments of the seed (Seon et al. 2002). The antibody is bound to the protein A-fused oleosin during seed grinding and can be recovered from oil bodies carrying the protein A and antibody complex using a simple extraction procedure (Seon et al. 2002). Another approach is so called ZERA technology (Eraplantech, www.eraplantech.com). Zera technology uses proline-rich (gamma)-zeins, which are maize storage proteins that accumulate inside large vesicles called protein bodies. When recombinant proteins fused to zeins are localized in endoplasmic reticulum, a high amount of the fusion proteins is accumulated to form stable bodies, which allows the production of recombinant proteins through their deposition on protein bodies of plant cells. Zera assemble heavy protein bodies in the endoplasmic reticulum (ER) of plant cells to highly accumulate and form supermolecular aggregates of polyproine structures and thus eases their recovery by simple homogenization and centrifugation. This technology has been successfully applied to produce calcitonin in tobacco plants. At present, ERA Plantech aims to validate and optimize ZERA technology to ensure its viability and competitiveness to produce a wide range of peptides and proteins such as antibodies.

Glycosylation

Antibodies are glycoproteins with specific glycoforms that are involved in their folding, stability, and activity. The altered glycosylation can influence biological characteristics such as antigen binding activity, affinity of Fc receptors, stability, and immunogenicity. Thus, the authentic *N*-glycosylation pattern on antibodies produced from any production system is necessary to obtain their intended therapeutic effect similar to the parental antibody. Plants have *N*-glycosylation capability similar to those of mammalian cells. However, *N*-glycosylation patterns processed in plant cells differ from those of mammals and humans (Fig. 6).

In plants, *N*-linked glycans contain β(1,2)-xlyose and α(1,3)-fucose instead of α(1,6)-fucose in mammals (Fig. 6). Furthermore, the plant *N*-glycan rarely carries galactose and lacks sialic acid (Fig. 6). These plant-specific glycans are considered to be potential antigenic and/or allergenic epitopes (Bakker et al. 2001). Although

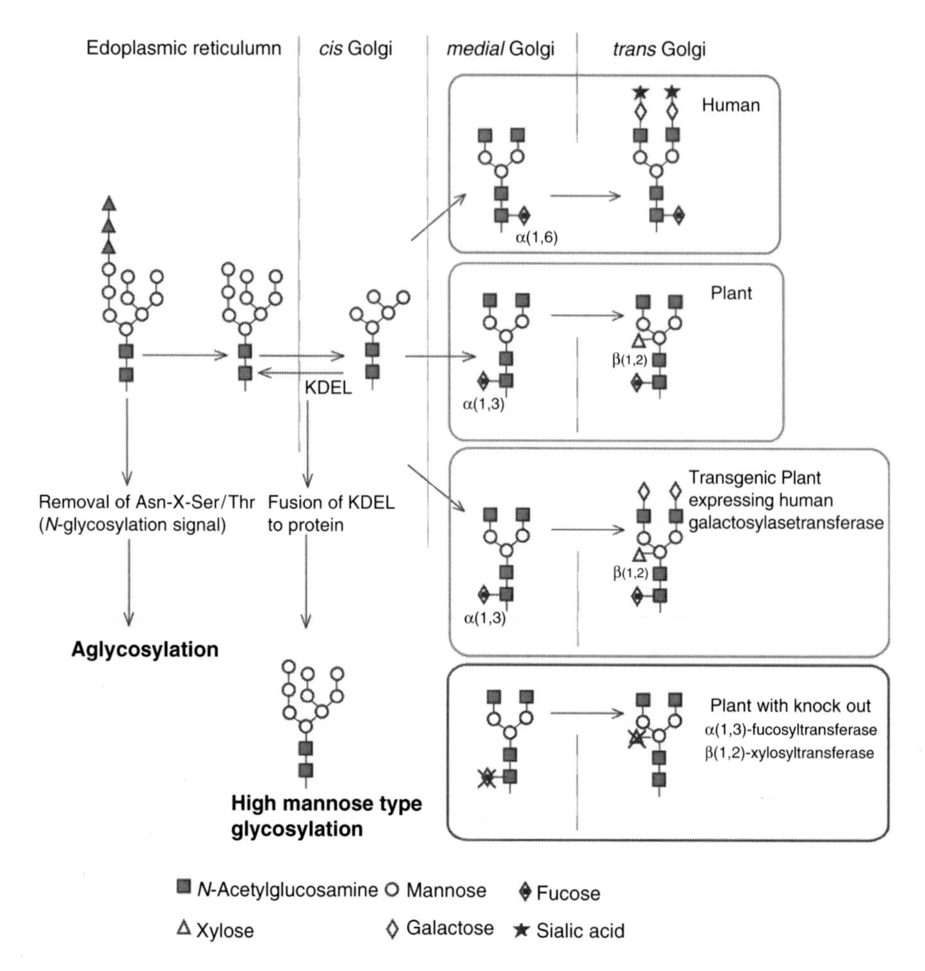

Fig. 6 *N*-glycosylation processing of glycoproteins in human and plant. *N*-glycosylation in the endoplasmic reticulum (ER) starts with transferring oligosaccharide precursor to Asn-X-Ser/Thr glycosylation signals on proteins. This precursor is processed by glycosidases and glycosyltransferases in the ER and the Golgi apparatus. Glycosylation processing in the ER is conserved in the plant and animal kingdoms and restricted to high mannose-type *N*-glycans, whereas the further glycosylation process in Golgi apparatus generates highly diverse matured *N*-glycan structures

plant-specific glycosylated antibody is not immunogenic in mice (Chargelegue et al. 2000), one of the major safety issues of plant-produced therapeutics remains these plant-specific *N*-glycosylation patterns. These potential concerns could become a drawback of the plant production system. Several research groups have studied removing plant-specific glycans, the potential antigenic components (Wenderoth and von Schaewen 2000; Ko et al. 2003; Finnem et al. 2005). One simple approach is aglycosylation to obtain antibodies with no *N*-glycosylation by mutating Asn-X-Ser/Thr sites, which are signals for *N*-glycosylation (Nuttall et al. 2005; Fig. 6). This approach is effective if the intended antibody activity is not affected by aglycosylation. Another approach is to retain the antibodies in ER to avoid plant-specific glycan residues such as β-(1,2)-xylose and α-(1,3)-fucose (Ko et al. 2003; Fig. 6). Glycosylation processing in the ER is conserved in the plant and animal kingdom and restricted to high mannose-type *N*-glycans, whereas the further glycosylation process in the Golgi apparatus where additional glycans are added for glycan maturation is highly diverse. When the ER-retention signal KDEL/HDEL fused to proteins, the signal retains proteins in the ER (Nuttall et al. 2002). We found that the KDEL fused to the heavy chains effectively generated 90% of the high mannose-type glycosylated antibodies in a total population of antibody produced in plants. The remaining 10% of the antibody population was GlcNAc2Man3GlcNAc2 (4.3%) and GlcNAc2Man3(xylose)GlcNAc2 (5.7%) without α(1,3)-fucose residues, which were often added on the trans-Golgi side (Fitchette-Laine et al. 1994; Ko et al. 2003). In this study, the KDEL signal fused to heavy chain does not completely retain antibodies to ER from Gogi, resulting in heterologous forms of glycan structures. Gomord et al. (2004) proposed fusion of KDEL to both light and heavy chains to increase high retrieval efficiency to obtain a homogenous population of high mannose-type glycan on full-size monoclonal antibody. When compared with mammal-derived antibodies, antibodies with high mannose-type glycans were rapidly cleared in vivo (Ko et al. 2003). This rapid clearance rate of antibodies may benefit rabies postexposure prophylaxis where vaccines are applied 10 days after passive immunization with antibodies. The dual effect of rabies postexposure treatment with both antibody and vaccine may often cause interference between passive and active immunization because of larger persistence of antibody in the circulation (Koprowski and Black 1952; Schumacher et al. 1992; Lang et al. 1998). Thus, the short half-life of antibodies may reduce potential interference between the two types of immunization. However, this strategy to generate high mannose-type glycosylated antibodies is not universally applicable because of its lower stability in vivo unless the high clearance rate is beneficial for therapeutic application.

Another approach to eliminate plant-specific glycan residues is to knock out the expression of glycosyltransferases involving β-(1,2)-xylosylation and α-(1,3)-fucosylation in Gogi apparatus enzymatic machinery (Gomord et al. 2004; Fig. 6). The *N*-glycan processing enzymes, β-(1,2)-xylosytransferase and α-(1,3)-fucosyltransferase, have been identified and cloned from different plant species. The expression of these enzymes has been inhibited using antisense technology. In moss (*Physcomitrella patens*), homologous recombination for gene targeting has been used to knock out expression of the α-(1,3)-fucosyltransferase and β-(1,2)-xylosyltransferase, eliminating plant-specific glycoepitopes without any effect on protein secretion (Koprivova et al. 2004). However, so far the approach has only been able to decrease the enzyme activities in *N*-glycan biosysnthesis in plants (Wenderoth and von Schaewen 2000).

In addition to eliminating plant-specific sugar, humanization of *N*-glycosylation is also essential for the production of authentic glycosylated antibodies in plants.

The strategy to humanize plant *N*-glycans is to express mammalian glycosyltransferases, which would complete *N*-glycan maturation in the plant's Golgi apparatus. The expression of human β(1,4)-galactosyltransferase in transgenic tobacco plants produces 30% galactosylated *N*-glycan in an antibody population, whose level is comparable to hybridoma cells (Bakker et al. 2001; Fig. 6). Addition of sialylation into plant *N*-glycosylation machinery is another important humanization process that can affect biological activity and in vivo half-life in the human body. Shah et al. (2003) reported the presence of sialylated glycoconjugates in suspension-cultured cells of *Arabidopsis thaliana* and suggested that a genetic and enzymatic basis for sialylation exists in plants. In contrast, Senveno et al. (2004) argued that sialylation does not exist in plants and proposed that obtaining sialylated glycan in plant cells requires the input and expression of at least five heterologous genes involving in entire sialylation pathway. Furthermore, these enzymes should be active and correctly targeted in Golgi compartment in plant cells. In nature, it seems that efficient sialylation processes are absent from plant cells. Thus, the humanization of antibodies with sialylation in plant systems, if necessary, remains a major challenge in producing authentic glycosylated antibodies in plant systems. New technology for the humanization of *N*-glycosylation will open up plant systems that efficiently produce authentic antibodies with enhanced bioactivities for human immunotherapy in the near future.

Conclusion

Plants have many advantages over other existing systems to produce monoclonal antibody. The advantages are recognized and appreciated with no doubt on the ability to produce functional antibodies, whereas the drawbacks related to the limited expression level, the authentic quality of antibody, downstream processing cost, public acceptance, and environmental concern hold their steps to utilization of plant systems as the commercial systems. A growing demand for therapeutic and diagnostic antibodies and the lack of current production facilities to meet the demand will be encountered. Thus, it is necessary to establish an alternative antibody production platform. New technologies are currently being developed to overcome the limitations of plant systems for antibody production.

Acknowledgments The authors thank Dr. Hilary Koprowski for his advice. This study was supported by a grant from Korea Research Foundation (KRF-2006-331-F00021) and the BioGreen 21 R&D Project of Rural Development Administration, Korea (20070410340026).

References

Bakker H, Bardor M, Molthoff JW, Gomord V, Elbers I, Stevens LH, Jordi W, Lommen A, Faye L, Lerouge P, Bosch D (2001) Galactose-extended glycans of antibodies produced by transgenic plants. Proc Natl Acad Sci USA 98:2899–2904

Baum TJ, Hiatt A, Parrott WA, Pratt LH, Hussey RS (1996) Expression in tobacco of a functional monoclonal antibody specific to stylet secretions of the root-knot nematode. Molecular Plant-Microbe Interaction 9:382–387

Boothe JG, Saponja JA, Parmenter DL (1997) Molecular farming in plants: oilseeds as vehicles for the production of pharmaceutical proteins. Drug Develop Res 42:172–181

Borisjuk NV, Borisjuk LG, Logendra S, Petersen F, Gleba Y, Raskin I (1999) Production of recombinant proteins in plant root exudates. Nat Biotechnol 17:466–469

Brockmann EC, Cooper M, Stromsten N, Vehniainen M, Saviranta P (2005) Selecting for antibody scFV fragments with improved stability using phage display with denaturation under reducing conditions. J Immunol Methods 296:159–170

Burton DR (2002) Antibodies, viruses and vaccines. Nat Rev Immunol 2:706–713

Burton DR, Barbas CFr (1994) Human antibodies from combinatorial libraries. Adv Immunol 57:191–280

Cabanes-Macheteau M, Fitchette-Laine AC, Loutelier-Bourhis C, Lange C, Vine ND, Ma JK, Lerouge P, Faye L (1999) N-Glycosylation of a mouse IgG expressed in transgenic tobacco plants. Glycobiology 9:365–372

Carter P (2001) Bispecific human IgG by design. J Immunol Methods 248:7–15

Casadevall A, Dadachova E, Pirofski L (2004) Passive antibody therapy for infectious diseases. Nat Microbiol 2:695–702

Chadd HE, Chamow SM (2001) Therapeutic antibody expression technology. Curr Opin Biotechnol 12:188–194

Chargelegue D, Vine ND, van Dolleweerd CJ, Drake PM, Ma JK (2000) A murine monoclonal antibody produced in transgenic plants with plant-specific glycans is not immunogenic in mice. Transgenic Res 9:187–194

Colby DW, Garg P, Holden T, Chao G, Webster JM, Messer A, Ingram VM, Wittrup KD (2004) Development of a human light chain variable domain (VL) intracellular antibody specific for the amino terminus of huntingtin via yeast surface display. J Mol Biol 342:901–912

Coloma MJ, Trinh RK, Martinez AR, Morrison SL (1999) Position effects of variable region carbohydrate on the affinity and in vivo behavior of an anti-(1-->6) dextran antibody. J Immunol 162:2162–2170

Conrad U, Fiedler U (1998) Compartment-specific accumulation of recombinant immunoglobulins in plant cells: an essential tool for antibody production and immunomodulation of physiological functions and pathogen activity. Plant Mol Biol 38:101–109

D'Aoust M-A, Lerouge P, Busse U, Bilodeau P, Trepanier S, Gomord V, Faye L, Vezina L-P (2004) Efficient and reliable production of pharmaceutical in alfalfa. In: Fischer R, Schillberg S (eds) Molecular Farming: Plant-made Pharmaceuticals and Technical Protein. Wiley, Weinheim, pp. 1–11

Daniell H (2002a) Medical molecular pharming: expression of antibodies, biopharmaceuticals and edible vaccines via the chloroplast genome. In: V. I.K (ed) Plant Biotechnology 2002 and Beyond, Plant Biotechnology 2002 and Beyond. Kluwer Academic Publishers, Orlando, pp. 371–376

Drake PMW, Chargelegue D, Vine ND, Van Dolleweerd CJ, Obregon P, Ma JK-C (2002) Transgenic plants expressing antibodies: a model for phytoremediation. FASEB J 16:1855–1860

Duering K (1988) Wundinduzierbare und Sekretion von T4 Lysozym und monoklonalen Antikorpern in Nicotiana tabcum. PhD thesis, Universität Koln, FRG

Fischer R, Schumann D, Zimmermann S, Drossard J, Sack M, Schillberg S (1999) Expression and characterization of bispecific single-chain Fv fragments produced in transgenic plants. Eur J Biochem 262:810–816

Fitchette-Laine AC, Gomord V, Chekkafi A, Faye L (1994) Distribution of xylosylation and fucosylation in the plant Golgi apparatus. Plant J 5:673–682

Gleba Y, Klimyuk V, Marillonnnet S (2005) Magnifection-a new platform for expressing recombinant vaccines in plants. Vaccine 23:2042–2048

Gomord V, Sourrouille C, Fitchette A-C, Bardor M, Pagny S, Lerouge P, Faye L (2004) Production and glycosylation of plant-made pharmaceuticals: the antibodies as a challenge. Plant Biotechnol J 2:83–100

Harris M (2004) Monoclonal antibodies as therapeutic agents for cancer. Lancet Oncol 5:292–302
Herlyn DM, Steplewski Z, Herlyn MF, Koprowski H (1980) Inhibition of growth of colorectal carcinoma in nude mice by monoclonal antibody. Cancer Res 40:717–721
Herlyn M, Steplewski Z, Herlyn D, Koprowski H (1986) CO17-1A and related monoclonal antibodies: Their production and characterization. Hybridoma 5:S3–S10
Hiatt A, Caffertey R, Bowdish K (1989) Production of antibodies in transgenic plants. Nature 342:76–78
Houghton AN, Scheinberg DA (2000) Monoclonal antibody therapies—a ' constant' threat to cancer. Nat Med 6:373–374
Hull AK, Criscuolo CJ, Mett V, Groen H, Steeman W, Westra H, Chapman G, Legutki B, Baillie L, Yusibov V (2005) Human-derived, plant-produced monoclonal antibody for the treatment of anthrax. Vaccine 23:2082–2086
Jobling SA, Jarman C, Teh MM, Holmberg N, Blake C, Verhoeyen ME (2003) Immunomodulation of enzyme function in plants by single-domain antibody fragments. Nat Biotechnol 21:77–80
Juwied M, Neumann R, Paik C, MJ, P-B, Sato J, van Osdol W, Weinstein JN (1992) Micropharmacology of monoclonal antibodies in solid tumors: direct experimental evidence for a binding site barrier. Cancer Res 52:5144–5153
Kathuria S, Sriraman R, Nath R, Sack M, Pal R, Artsaenko O, Talwar GP, Fischer R, Finnern R (2002) Efficacy of plant-produced recombinant antibodies against HCG. Hum Reprod 17:2054–2061
Ko K, Koprowski, H (2005) Plant biopharming of monoclonal antibodies. Virus Res 111:93–100
Ko K, Steplewski Z, Glogowska M, Koprowski H (2005) Inhibition of tumor growth by plant-derived mAb. Proc Natl Acad Sci USA 102:7026–7030
Ko K, Tekoah Y, Rudd PM, Harvey DJ, Dwek RA, Spitsin S, Hanlon CA, Rupprecht C, Dietzschold B, Golovkin M, Koprowski H (2003) Function and glycosylation of plant-derived antiviral monoclonal antibody. Proc Natl Acad Sci USA 100:8013–8018
Kohler G, Milstein C (1975) Continuous cultures of fused cells secreting antibody of predefined specificity. Nature 256:495–497
Koprivova A, Stemmer C, Altmann F, Hoffmann A, Kopriva S, Gorr G, Reski R, Decker EL (2004) Targeted knockouts of Physcomitrella lacking plant-specific immunogenic N-glycans. Plant Biotechnol J 2:517–523
Koprowski H, Black J (1952) Studies on chick-embryo-adapted rabies virus. J Immunol 72:79–84
Koprowski H, Yusibov V (2001) The green revolution: plants as heterologous expression vectors. Vaccine 19:2735–2741
Lang J, Simanjuntak GH, Soerjosembodo S, Koesharyono C (1998) Suppressant effect of human or equine rabies immunoglobulins on the immunogenicity of post-exposure rabies vaccination under the 2-1-1 regimen: a field trial in Indonesia. MAS054 Clinical Investigator Group. Bull World Health Organ 76:491–495
Le Gall F, Bove J-M, Garnier M (1998) Engineering of a single-chain variable-fragment (scFv) antibody specific for the stolbur phytoplasma (mollicute) and its expression in Escherichia coli and tobacco plants. Appl Environ Microbiol 64:4566–4572
Ma JK, Drake PM, Christou P (2003) The production of recombinant pharmaceutical proteins in plants. Nat Rev Genet 4:794–805
Ma JK, Hiatt A, Hein M, Vine ND, Wang F, Stabila P, van Dolleweerd C, Mostov K, Lehner T (1995) Generation and assembly of secretory antibodies in plants. Science 268:716–719
Malhotra R, Wormald MR, Rudd PM, Fischer PB, Dwek RA, Sim RB (1995) Glycosylation changes of IgG associated with rheumatoid arthritis can activate complement via the mannose-binding protein. Nat Med 1:237–243
Marillonnnet S, Thoeringer C, Kandzia R, Klimyuk V, Gleba Y (2005) Systemic Agrobacterium tumefaciens-mediated transfection of viral replicons for efficient transient expression in plants. Nat Biotechnol 23:718–723
Matusuuchi L, Sharon J, Morrison SL (1981) An analysis of heavy chain glycospeptides of hybridoma antibodies: correlation between antibody specificity and sialic acid content. J Immunol 127:2188–2190

Mayfield SP, Franklin SE, Lerner RA (2003) Expression and assembly of a fully active antibody in algae. Proc Natl Acad Sci USA 100:438–442

McCall AM, Shahied L, Amoroso AR, Horak EM, Simmons HH, Nielson U, Adams GP, Schier R, Marks JD, Wiener LM (2001) Increasing the affinity for tumor antigen enhances bispecific antibody cytotoxicity. J Immunol 166:112–117

McCormick AA, Kumagai MH, Hanley K, Turpen TH, Hakim I, Grill LK, Tuse D, Levy S, Levy R (1999) Rapid production of specific vaccines for lymphoma by expression of the tumor-derived single-chain Fv epitopes in tobacco. Proc Natl Acad Sci USA 96:703–708

Miroshinichenko S, Tripp J, Nieden U, Neumann D, Conrad U, Manteuffel R (2005) Immunomodulation of function of small heat shock proteins prevents their assembly into heat stress granules and results in cell death at sublethal temperatures. Plant J 41:269–281

Montefiori DC (2005) Neutralizing antibodies take a swipe at HIV in vivo. Nat Med 11:593–594

Nosanchuk JD, Steenbergen JN, Shi L, Deepe GSJ, Casadevall A (2003) Antibodies to a cell surface histone-like protein protect against Histoplasma capsulatum. J Clin Invest 112:1164–1175

Nowakowski A, Wang C, Powers DB, Amersdorfer P, Smith TJ, Montgomery VA, Sheridan R, Blake R, Smith LA, Marks JD (2002) Potent neutralization of botulinum neurotoxin by recombinant oligoclonal antibody. Proc Natl Acad Sci USA 99:11346–11350

Nuttall J, Ma JK-C, Frigerio L (2005) A functional antibody lacking N-linked glycans is efficiently folded, assembled and secreted by tobacco mesophyll protoplasts. Plant Biotechnol J 3:497–504

Nuttall J, Vine ND, Hadlington JL, Drake P, Frigerio L, Ma JK-C (2002) ER-resident chaperone interactions with recombinant antibodies in transgenic plants. FEBS 269:6042–6051

Peeters K, De Wilde C, Depicker A (2001) Highly efficient targeting and accumulation of a F(ab) fragment within the secretory pathway and apoplast of Arabidopsis thaliana. Eur J Biochem 268:4251–4260

Prosniak M, Faber M, Hanlon CA, Rupprecht C, Hooper DC, Dietzschold B (2003) Development of a cocktail of recombinant-expressed human rabies virus-neutralizing monoclonal antibodies for post-exposure prophylaxis of rabies. J Infect Dis 187:30386–30389

Schier R, Balint RF, McCall A, Apell G, Larrick JW, Marks JD (1996) Identification of functional and structural amino-acid residues by parsimonious mutagenesis. Gene 169:147–155

Schiermeyer A, Dorfmuller S, Schinkel H (2004) Production of pharmaceutical proteins in plants and plant cell suspension cultures.

Schillberg S, Zimmermann S, Zhan M-Y, Fischer R (2001) Antibody-based resistance to plant pathogens. Transgenic Res 10:1–12

Schlesinger JJ, Foltzer M, Chapman S (1993) The Fc portion of antibody to yellow fever virus NS1 is a determinant of protection against YF encephalitis in mice. Virology 192:132–141

Schumacher CL, Ertl HC, Koprowski H, Dietzschold B (1992) Inhibition of immune responses against rabies virus by monoclonal antibodies directed against rabies virus antigens. Vaccine 10:754–760

Senveno M, Bardor M, Paccalet T (2004) Glycoprotein sialylation in plants? Nat Biotechnol 22:1351–1352

Seon JH, Szarka S, Moloney M (2002) A unique strategy for recovering recombinant proteins from molecular farming: affinity capture on engineered oilbodies. J Plant Biotechnol 4:95–101

Shah MM, Fujiyama K, Flynn CR, Joshi L (2003) Presence of sialylated endogenous glyconjugates in plant cells. Nat Biotechnol 21:1470–1471

Shields RL, Lai J, Keck, R, O'Connell, L.Y, Hong, K, Meng, Y.G, Weikert, S.H.A, Presta, L (2002) Lack of Fucose on human IgG1 N-linked Oligosaccharide improves binding to human FcgRIII and antibody-dependent cellular toxicity. J Biol Chem 277:26733–26740

Smith G, Walmsley A, Polkinghorne I (1997) Plant-derived immunocontraceptive vaccines. Reprod Fertil Dev 9:85–89

Song WY, Sohn EJ, Martinoia E, Lee YJ, Yang YY, Jasinski M, Forestier C, Hwang I, Lee Y (2003) Engineering tolerance and accumulation of lead and cadmium in transgenic plants. Nat Biotechnol 21:914–919

Stoger E, Sack M, Fischer R, Christou P (2002) Plantibodies: applications, advantages and bottlenecks. Curr Opin Biotechnol 13:161–166

Stoger E, Sack M, Nicholson L, Fischer R, Christou P (2005) Recent progress in plantibody technology. Curr Pharmaceut Des 11:2439–2457

Stoger E, Vaquero C, Torres E, Sack M, Nicholson L, Drossard J, Williams S, Keen D, Perrin Y, Christou P, Fischer R (2000) Cereal crops as viable production and storage systems for pharmaceutical scFv antibodies. Plant Mol Biol 42:583–590

Tachibana H, Shirahata S, Murakami H (1992) Generation of specificity-variant antibodies by alteration of carbohydrate in light chain of human monoclonal antibodies. Biochem Biophys Res Commun 189:625–632

Trkola A, Kuster H, Rusert P, Joos B, Fischer M, Leemann C, Manrique A, Huber M, Rehr M, Oxenius A, Weber R, Stiegler G, Vcelar B, Katinger H, Aceto L, Gunthard HF (2005) Delay of HIV-1 rebound after cessation of antiretoviral therapy through passive transfer of human neutralizing antibodies. Nat Med 11:615–622

Umana P, Jean-Mairet J, Moudry R, Amstutz H, Bailey JE (1999) Engineered glycoforms of an antineuro-blastoma IgG1 with optimized antibody-dependent cellular cytotoxic activity. Nat Biotechnol 17:176–180

Vaccaro C, Zhou J, Ober RJ, Ward E.S (2005) Engineering the Fc region of immunoglobulin G to modulate in vivo antibody levels. Nat Biotechnol 23:1283–1288

Valdes R, Gomez L, Padilla S, Brito J, Reyes B, Alvarez T, Mendoza O, Herrera O, Ferro W, Pujol M, Leal V, Linares M, Hevia Y, Garcia C, Mila L, Garcia O, Sanchez R, Acosta A, Geada D, Paez R, Luis Vega J, Borroto C (2003) Large-scale purification of an antibody directed against hepatitis B surface antigen from transgenic tobacco plants. Biochem Biophys Res Commun 308:94–100

Vaquero C, Sack M, Schuster F, Finnern R, Drossard J, Schumann D, Reimann A, Fischer R (2002) A carcinoembryonic antigen-specific diabody produced in tobacco. FASEB J 16:408–410

Verch T, Yusibov V, Koprowski H (1998) Expression and assembly of a full-length monoclonal antibody in plants using a plant virus vector. J Immunological Methods 220:69–75

Vietta ES, Uhr JW (1994) Monoclonal antibodies as agonists: an expanded role for their use in cancer therapy. Cancer Res 54:5301–5309

Voss A, Niersbach M, Hain R, Hirsch HJ, Liao YC, Kreuzaler F, Fischer R (1995) Reduced virus infectivity in *N. tabacum* secreting a TMV-specific full-size antibody. Molecular Breeding 1:39–50

Vrain TC (1999) Engineering natural and synthetic resistance for nematode management. J Nematol 31:424–436

Vuist WM, Levy R, Maloney DG (1994) Lymphoma regression induced by monoclonal anti-idiotypic antibodies correlates with their ability to induce Ig signal transduction and is not prevented by tumor expression of high levels of bcl-2 protein. Blood 83:899–906

Weiner LM, Carter P (2005) Tunable antibodies. Nat Biotechnol 23:556–557

Wenderoth I, von Schaewen A (2000) Isolation and characterization of plant- N-acetyl glucosaminyltransferase I (GntI) cDNA sequences. Functional analyses in the Arabidopsis cgl mutant and in antisense plants. Plant Physiol 123:1097–1108

Winter G, Griffiths AD, Hawkins RE, Hoogenboom HR (1994) Making antibodies by phage display technology. Annu Rev Immunol 12:433–455

Wright A, Morrison SL (1998) Effect of C2-associated carbohydrate structure on Ig effector function: studies with chimeric mouse-human IgG1 antibodies in glycosylation mutants of Chinese hamster ovary cells. J Immunol 160:3393–3402

Wright A, Tao MH, Kabat EA, Morrison SL (1991) Antibody variable region glycosylation: position effects on antigen binding and carbohydrate structure. EMBO J 10:2717–2123

Plant Production of Veterinary Vaccines and Therapeutics

R.W. Hammond and L.G. Nemchinov

Contents

Abstract Plant-derived biologicals for use in animal health are becoming an increasingly important target for research into alternative, improved methods for disease control. Although there are no commercial products on the market yet, the development and testing of oral, plant-based vaccines is now beyond the proof-of-principle stage. Vaccines, such as those developed for porcine transmissible gastroenteritis virus, have the potential to stimulate both mucosal and systemic, as well as, lactogenic immunity as has already been seen in target animal trials. Plants

R.W. Hammond(✉) and L.G. Nemchinov
USDA-ARS Molecular Plant Pathology Laboratory, Beltsville, MD 20705, USA
e.mal: rose.hammond@ars.usda.gov

A.V. Karasev (ed.) *Plant-produced Microbial Vaccines*.
Current Topics in Microbiology and Immunology 332

are a promising production system, but they must compete with existing vaccines and protein production platforms. In addition, regulatory hurdles will need to be overcome, and industry and public acceptance of the technology are important in establishing successful products.

Introduction

Veterinary pharmaceutical products generated US $14.5 billion in worldwide sales in 2000, the leading biological products of which were vaccines against foot-and-mouth disease (Gay et al. 2003). For viral diseases, the most common antigen and route of vaccination are live attenuated or chemically inactivated whole viruses via parenteral injection. For diseases caused by bacteria, live attenuated or killed bacteria or outer membrane protein preparations elicit protective antibody responses when injected, given orally, or given intranasally (Bowersock and Martin 1999). Recent developments of subunit or peptide vaccines by incorporation of known protective antigenic proteins and/or serologically dominant epitopes into DNA plasmids or mammalian viral-based vectors for delivery have prompted investigation into the use of plant expression systems for production of candidate animal vaccines (Streatfield and Howard 2003a). In addition to the safety advantages through the reduced risk of contamination with animal or human pathogens, plant-based vaccines also offer the opportunity to deliver the vaccine cost-effectively and orally to large numbers of animals at the same time, thus saving time and the cost of immunizing animals individually by injection, as well as increasing client compliance, thereby contributing to overall herd health (Streatfield et al. 2001). There are new growth opportunities in veterinary applications of plant-based biologics, including vaccines and therapeutic products.

This chapter reviews the current status of the development of plant-derived biologics for administration to animals and for use as diagnostic tools. Development of stable, transgenic plants and plant viral-based vectors for transient expression of biopharmaceuticals for control of animal diseases has been carried out, with either full-length proteins or peptides, showing efficacy in animal trials.

Foot-and-Mouth Disease Virus

Foot-and-mouth disease virus (FMDV) is the causative agent of an economically important disease that affects meat-producing animals, including cattle, sheep, pigs, and other wild and domestic cloven-hoofed animals, and remains one of the most important pathogens of livestock (Mason et al. 2003). The disease, which is acute and highly contagious, is spread by contact with infected animals, and by movement of contaminated vehicles, humans, and nonsusceptible animals.

Current control measures include restrictions on the movement of animals and animal products, slaughter of affected animals, disinfection, and vaccination with chemically inactivated tissue culture-propagated FMDV (Mason et al. 2003); these vaccines require high-containment facilities for vaccine production.

FMDV is a member of the genus *Aphthovirus* of the family Picornaviridae. It is a nonenveloped, isometric virus that contains four capsid proteins (VP1–4; also known as 1D, 1B, 1C, and 1A, respectively) and a single molecule of polyadenylated RNA that is translated into a single polyprotein that undergoes proteolytic cleavage to mature, functional proteins (Levy et al. 1994). Studies have shown that VP1 carries epitopes responsible for induction of neutralizing antibodies. Immunization with VP1, synthetic peptides derived from VP1, or recombinant vaccinia virus expressing VP1, induces protection against challenge virus in natural and experimental hosts (Brown 1992; Berinstein et al. 2000, respectively).

Strategies that have been developed for the expression of FMDV antigens in plants include fusion of immunodominant epitopes of FMDV capsid protein VP1 with plant virus capsids. In addition, expression of the complete VP1 structural protein by stable incorporation in transgenic plants or by expression from a viral-based vector have been reported. These strategies have yielded promising results, which are summarized below.

The earliest attempt at producing a plant-derived vaccine to FMDV was by fusion of an immunodominant peptide epitope [amino acids (aa) 141–160 of the 211-aa protein] of VP1 from the O serotype into the βB–βC loop of cowpea mosaic virus (CPMV) S capsid protein (Usha et al. 1993), thereby incorporating the epitope into virus particles. Although CPMV containing the inserted epitope did not spread systemically in infected plants, the chimeric capsid protein isolated from plant protoplasts or inoculated leaves could be detected on Western blots using antiserum raised against the free homologous peptide. Since that time, optimization of the design of CPMV chimeric proteins has resulted in the expression of several FMDV epitopes (Porta et al. 2003), thereby demonstrating that the plant viral CP could be used as a presentation system for FMDV epitopes.

A tobacco mosaic virus (TMV)-based vector was also employed to express similar immunogenically dominant FMDV VP1 epitopes as a fusion with the TMV capsid protein (Wu et al. 2003). TMV recombinants, expressing either an 11- or 14-aa epitope of VP1, systemically infected tobacco plants, and the epitopes were stably expressed in the virus particles. The authors estimated that 0.3–0.4 g of the FMDV epitope was expressed per kilogram of infected leaf tissue. Guinea pigs either immunized parenterally or by oral administration of purified virus were protected at some level against an FMDV challenge given at 42 days after the first immunization dose. Although the oral presentation had some protective effects, it was less effective than parenteral administration.

TMV was also employed earlier to express the VP1 (serotype 01 Campos, 01C) complete structural protein from a separate mRNA utilizing the duplicate subgenomic promoter of the virus-based vector (Wigdorovitz et al. 1999b). *Nicotiana benthamiana* plants infected with the recombinant virus produced the protein in a soluble form at 50–150 μg per gram of freshly harvested leaves, although there was some evidence of proteolytic degradation. Mice parenterally immunized with crude leaf extracts were protected from FMDV challenge.

Alternative strategies for expression of FMDV antigens involved the development of transgenic plants where the foreign gene was stably integrated into the plant genome. The complete FMDV VP1 structural protein [serotype O1 Campos (O1C); ~23 kDa] was expressed in transgenic plants under control of the CaMV 35S promoter in *Arabidopsis* (Carrillo et al. 1998), alfalfa (Wigdorovitz et al. 1999a), and potato (Carrillo et al. 2001). Mice injected with leaf extracts from *Arabidopsis* plants expressing the VP1 protein were protected from FMDV challenge inoculation (Carillo et al. 1998). Although the levels of VP1 expression were low, mice either injected with leaf extracts of transgenic alfalfa or fed fresh transgenic leaf material were protected against FMDV challenge (Wigdorovitz et al. 1999a). In transgenic potato leaves, estimates of VP1 expression were between 0.005% and 0.01% of the total soluble protein (TSP). Mice immunized intraperitoneally with foliar extracts of transgenic potato plants displayed an immune response to the recombinant proteins as well as to purified FMDV particles (Carillo et al. 2001). When the immunized mice were experimentally challenged with infectious FMDV by intraperitoneal inoculation, 90% of the vaccinated mice were protected against infection as compared to the control group (Wigdorovitz et al. 2004).

The low expression levels of VP1 in transgenic plants found in earlier studies prompted Dus Santos et al. (2002) to investigate the possibility of increasing the vaccine epitope concentration by fusing the protective epitope (aa 135–160) of VP1 to the amino terminus of glucuronidase (gus A reporter gene). Under transcriptional control of the CaMV 35S promoter, protein levels of 0.5–1 mg per gram of TSP were realized (0.005%–0.01%) in alfalfa. These levels are similar to those found by Carillo et al. (2001) in transgenic potato. Crude extracts were used to immunize mice by intraperitoneal inoculation and there was complete protection against FMDV challenge.

In a report by Sun et al. (2003), accumulation of up to 3% TSP was obtained for a fusion of VP1 with the cholera toxin B subunit (CTB) and transformation of the green alga, *Chlamydomonas reinhardtii*, chloroplast genome. The fusion protein retained VP1 antigenicity, but no animal trials were conducted.

Finally, Dus Santos et al. (2005) reported the development of transgenic alfalfa plants expressing the complete FMDV structural polyprotein gene P1 and the 3C protease as experimental immunogens. A DNA fragment encoding the P1–3C protein was engineered under the control of the CaMV35S promoter. Foliar extracts of transgenic plant tissue elicited an anti-FMDV immune response in mice inoculated intraperitoneally. Vaccinated mice exhibiting an immune response and experimentally challenged with infectious FMDV were protected against the challenge infection.

The research described above demonstrates the feasibility of a plant virus-based vaccine against FMDV and indicates that its commercialization could be possible in the near future.

Transmissible Gastroenteritis Virus

Swine transmissible gastroenteritis virus (TGEV) is the causative agent of a highly contagious, severe, acute diarrhea of newborn piglets, resulting in high mortality rates of piglets under 2 weeks of age. Protective immunity against this disease must be developed in pregnant, TGEV-immune sows so that passive protection can be passed (mainly in the form of IgA antibodies capable of surviving the gastrointestinal tract) to piglets through colostrum and milk (Saif et al. 1994). Pigs that survive first infection are immune from subsequent infections. A commercially available modified live vaccine is used to control the disease.

TGEV, a member of the *Coronavirus* genus of the family Coronaviridae, is an enveloped, single-stranded RNA virus. It contains three structural proteins, M, N, and S. M is an integral membrane protein, N, a phosphoprotein that encapsulates the RNA, and S, the surface or spike glycoprotein (gS) (Laude et al. 1990). Neutralizing antibodies against the virus are directed mainly to the gS protein, and neutralizing epitopes have been mapped to the N-terminal domain of this protein. Four major antigenic sites have been mapped on the gS protein of which site A is immunodominant. The gS protein has been expressed in different mammalian expression vectors, including adenovirus, which promoted systemic and mucosal immunity and conferred protection to suckling piglets (Torres et al. 1996).

Transgenic plants have been developed to express the entire gS protein or portions thereof as candidate plant-based vaccines. Gomez et al. (1998) transformed *Arabidopsis* with cDNA constructs containing the N-terminal domain (aa 1–750) or the full-length gS protein under transcriptional control of the CaMV 35S promoter. Mice inoculated intramuscularly with leaf extracts containing 0.03%–0.06% TSP of gS from transgenic plants developed TGEV-specific antibodies which immunoprecipitated the virus protein and neutralized virus infectivity in vitro. Transgenic potato plants were also created that expressed the TGEV N-terminal domain of the glycoprotein S (N-gS), containing the major antigenic sites of the protein. Levels of expression, 0.02%–0.05% TSP of potato tubers, were similar to those achieved in *Arabidopsis*. Extracts of potato tubers expressing the protein were inoculated intraperitoneally to mice, and the vaccinated mice developed serum IgG specific for TGEV. When potato tubers expressing N-gS were fed directly to mice, they developed serum antibodies specific for gS protein, demonstrating oral immunogenicity of the plant-derived spike protein (Gomez et al. 2000). Similarly, antigens purified from transgenic tobacco plants expressing three different regions of the TGEV gS protein-induced TGEV-specific immune responses in pigs as determined by virus neutralization (Tuboly et al. 2000).

A corn-based production system has also been developed, primarily for the delivery of oral vacccines to TGEV (Lamphear et al. 2002, 2004; Streatfield et al. 2001, 2002). A synthetic, maize codon-optimized version of the TGEV gS protein was expressed in-frame with a maize codon-optimized version of the signal sequence for cell secretion from the barley α-amylase protein (Streatfield et al. 2001). Expression levels of up to 2% TSP were observed in transgenic corn seed and the protein was shown to assemble into the active pentameric form in planta (Streatfield et al. 2002). When corn seed expressing the antigen was fed to piglets, partial protection against subsequent TGEV challenge was induced. Although TGEV-corn-fed piglets showed fewer overall symptoms compared to control unvaccinated piglets, they recovered more slowly from infection than piglets vaccinated with modified live virus.

Commercial processing methods were tested in order to examine the distribution of antigen in the major seed compartments—germ, grits, and bran (Lamphear et al. 2002). TGEV-S antigen was enriched in the germ fraction. Storage of the germ meal at 4°C or whole seeds at higher temperatures for up to 1 year had negligible effects on antigen levels. These results showed that the antigen could survive standard grain processing. Furthermore, the antigen could be enriched in a particular fraction (up to 500 μg/g of germ), thereby reducing the volume necessary per dose of vaccine and allowing it to be easily incorporated into animal feed.

Subsequent studies of prime/boost vaccination and efficacy of the corn-derived oral vaccine showed that rapid induction of neutralizing antibodies occurred in piglets, which previously did not contain detectable neutralizing antibodies in the serum. This suggests that there was a memory response to previous oral ingestion of the antigen (Lamphear et al. 2002). The swine feeding studies were extended to evaluate the ability of the corn-based oral vaccine to boost immune responses in young sows previously sensitized with the commercially available modified live viral vaccines (Lamphear et al. 2004). It was found that the oral immunization could boost neutralizing antibody levels in the young sows in the serum, colostrum, and milk (lactogenic immunity), with the potential to confer protection to suckling piglets. In light of the results obtained by Walmsley et al. (2003) showing the passive immunization of mice pups through oral immunization of the mother, commercial application of a plant-based oral vaccine as a boost to large herds of swine shows great potential.

Canine Parvovirus and Mink Enteritis Virus

Canine parvovirus (CPV) and mink enteritis virus (MEV) are members of the *Parvovirus* genus of the family Parvoviridae, which are nonenveloped, single-stranded DNA viruses (Levy et al. 1994). The virus possesses three structural proteins, VP1, VP2 (the major structural protein of CPV), and VP3. VP1 and VP2 are splicing products from the same gene. Only VP2 is required for viral particle formation. CPV, MEV, and feline panleukopenia virus (FPLV) are host-specific variants of CPV (sharing 98% amino acid sequence homology) that can infect dogs, mink, cats, or raccoons. These viruses are of great economic importance and clinical manifestations of the disease include diarrhea, severe inflammation of the intestine, and anorexia. In cats, FPLV induces a severe febrile disease. The disease can be controlled by vaccination with chemically inactivated or live, attenuated virus.

It is known that protective antibodies produced during CPV infection are specific for the structural proteins. Extensive studies of surface epitopes of VP2 and synthetic peptide vaccines against CPV based on the the surface epitopes have been described. Among these, the amino terminus of VP2 contains peptides (linear peptide epitopes) that have been shown to induce neutralizing activities against CPV in mice, rabbits, and mink (Langeveld et al. 1994, 1995; Casal et al. 1995).

The development of antigen presentation systems in plants for creation of CPV and related candidate vaccines has been primarily based on the fusion of the parvovirus neutralizing epitopes, shown previously to provide full protection against challenge infection, with plant virus capsid proteins. Dalsgaard et al. (1997) first reported the fusion of peptide 3L17 (aa 3–21 of VP2; DGAVQPDGGQPAVRNER) to the capsid protein of CPMV. The chimeric virus particles were produced by infection of black-eyed bean and 50–60 mg virus could be purified from 50 g of leaf material. Subcutaneous injection of mink with purified virus particles conferred protection against clinical disease and abolished shedding of virus after challenge with virulent MEV (Dalsgaard et al. 1997). This was the first demonstration that an experimental vaccine produced in plants was able to confer protection against infectious disease challenge in an animal. The reduction or abolishment of virus shedding is an important step toward containing the virus, as it is often difficult to isolate infected animals to prevent spread to other animals in a hospital, home, pet store, or dog parks.

In a separate study, Fernández-Fernández et al. (1998), fused a peptide corresponding to aa 2–21 (2L21) of VP2 to the amino terminus of the plum pox potyvirus capsid protein. Antigenicity of the chimeric coat protein virus was demonstrated by immunization of mice and rabbits, and although the antibodies showed neutralizing activity, the antibody titers were very low.

An alternative expression strategy was reported by Gil et al. (2001), who described the high-yield production of a CPV peptide vaccine in transgenic plants. A 21-mer linear antigenic peptide (2L21) was expressed as an amino terminal fusion with the GUS protein and under transcriptional control of the CaMV 35S promoter. Expression levels of the recombinant protein in *Arabidopsis thaliana* were up to 3% TSP, a production yield comparable to that obtained with the same epitope expressed by chimeric viruses. The immunogenicity of the plant-derived peptide was demonstrated in mice immunized either intraperitoneally or orally with transgenic plant extracts.

Finally, in follow-up studies using chimeric plant virus particles as immunogens for CPV, Langeveld et al. (2001) and Nicholas et al. (2002, 2003), report improved safety/delivery and immunization protocols, respectively. CPMV particles expressing the 3L17 linear epitopeof VP2 (Dalsgaard et al. 1997) were inactivated with UV light to remove the possibility of replication of the purified virus in a plant host after manufacture of the vaccine (Langeveld et al. 2001). Parenteral vaccination with the inactivated virus was able to protect dogs from a lethal challenge with CPV. The dogs displayed no signs of clinical disease, did not shed CPV particles, and had high titers of peptide-specific neutralizing antibodies. Using the same experimental immunogen, combinations of systemic and mucosal routes for priming and boosting immunizations were tested for their influence on the immune response in mice (Nicholas et al. 2003). Serum antibody responses were greatest in animals receiving subcutaneous prime and boosting; the response was least in mucosally vaccinated animals. The route of administration did not alter antibody ratios; intranasal administration following subcutaneous priming was effective in inducing mucosal IgA responses. These studies have implications for the development of effective immunization strategies using chimeric virus particles for protection against mucosally acquired viral infections.

Rabies Virus

Rabies virus, a member of the genus *Lyssavirus* of the family Rhabdoviridae, causes an acute and deadly viral infection of the central nervous system. It remains a significant threat to human and animal health. Although rabies in humans is rare in the United States, as many as 18,000 Americans get rabies shots each year because they have been in contact with animals that may be rabid. In 1998, according to the United States Centers for Disease Control and Prevention (CDC), only one person died of rabies in this country. In other parts of the world, however, many people die of rabies each year. The World Health Organization (WHO) estimates that around the world more than 40,000 people die every year from rabies. WHO also estimates that 10 million people worldwide are treated after being exposed to animals that may have rabies (http://www.niaid.nih.gov/factsheets/rabies.htm). The virus, which is in the saliva of infected animals, is usually transmitted by bites from infected animals. All warm-blooded animals can get rabies, and some may serve as natural reservoirs of the virus. Therefore, control of the infection in wild animals will reduce the risk of infection in humans and domesticated animals.

The virion of rabies virus consists of a lipid-rich envelope that covers a helical ribonucleocapsid core consisting of a negative-sense RNA genome (Levy et al. 1994). The genome encodes five proteins, one of which is a glycoprotein (G) that forms approximately 400 spikes that are tightly arranged on the surface of the particle. Fusion of the rabies virus particle to the host cell membrane initiates the infection process, which may involve interaction of the G proteins with specific cell surface receptors The G protein possesses hemagglutinin activity and is the target of neutralizing antibodies. Oral immunization with bait containing a vaccinia virus-rabies glycoprotein recombinant has been shown to protect raccoons and foxes against the disease (Rupprecht et al. 1986; Brochier et al. 1990, respectively).

Efforts have been made to develop a cheap, safe, oral plant-based vaccine to control rabies in animals and humans. McGarvey et al. (1995) engineered tomato plants to express the viral G protein. The protein, which was expressed in leaf and fruit tissues, was immunoreactive with anti-G antibodies. Yusibov et al. (1997) expressed a B cell epitope from rabies glycoprotein (G) G5–24 and a T cell epitope from rabies nucleoprotein (N) 31D as chimeric protein fusions with the N terminus of the coat protein of the plant virus Alfalfa mosaic virus (AMV). This allowed the recombinant epitopes to be displayed on the surface of spherical AMV particles.

The chimeric virus proteins were translated from subgenomic RNAs expressed from a TMV-based plant virus expression vector. AMV virus-like particles (VLPs) were purified from infected plant tissue and used to immunize mice intraperitoneally. High titers of rabies-specific antibodies were detected in mice immunized with the purified particles and sera from the immunized mice neutralized the CVS-11 strain of rabies virus in vitro.

In a follow-up study, Modelska et al. (1998) reported on the development of local and systemic immune responses in mice, immunized either intraperitoneally or orally by feeding with virus-infected spinach leaves containing the engineered virus displaying the rabies antigen. Forty percent of the intraperitoneally immunized mice were protected against challenge with a lethal dose of rabies virus; orally immunized mice (feeding with raw, virus-infected spinach leaves or by gastric intubation) showed stimulation of both IgG and IgA production and weakened signs of the disease.

Yusibov et al. (2002) reported that recombinant virus or VLPs expressing the G and N protein epitopes, produced using two additional expression strategies, protected mice from challenge infection with rabies when they were immunized parenterally. In addition, virus-infected, unprocessed, raw spinach leaves were orally administered to human volunteers who had either been preimmunized with a conventional rabies vaccine or had not been immunized. Those who had been previously vaccinated showed a response against the peptide antigen after ingesting the spinach leaves. Five of nine individuals who had not been previously immunized demonstrated significant antibody responses to the antigen, and, when given a dose of conventional vaccine, three of the individuals showed detectable levels of rabies virus-neutralizing antibodies.

These results demonstrate that recombinant, subunit-based, plant-manufactured rabies vaccines can be delivered by injection or orally and show promise for veterinary applications. For example, raccoon rabies is common in parts of the Eastern United States. Raccoon rabies spreads rapidly and infects large numbers of raccoons. The disease often spreads to other wildlife and pets, making human exposure a real concern. To address this problem, an oral vaccination program using recombinant vaccinia virus containing the rabies antigen is incorporated into bait. A plant-derived vaccine could easily be incorporated into a bait. Management of rabies in other animals, including dogs, cats, ferrets, and livestock, involves parenteral injection of inactivated virus, necessitating the production of large quantities of rabies virus in cell culture. A plant-based subunit vaccine would eliminate the risk of exposure to vaccine producers. It would also eliminate the chance of infection in the vaccinated animal if the inactivated virus vaccine retains some viability.

Rabbit Hemorrhagic Fever Virus

Rabbit hemorrhagic disease is a rapidly spreading, lethal infection of adult animals in the wild rabbit population and affected farms. Infected rabbits usually die within 48–72 h of necrotizing hepatitis. Current vaccines are based on formalin-inactivated liver homogenates of infected animals (Peeters et al. 1992).

The causal agent of the disease, rabbit hemorrhagic disease virus (RHDV), is a member of the family Caliciviridae, that also includes feline calicivirus, Norwalk virus, and swine vesicular exanthema virus. It is nonenveloped virus with a single-stranded RNA genome that contains one long open reading frame encoding structural and nonstructural proteins (Parra and Prieto 1990). The major RHDV structural component of the capsid is a 60-kDa protein known as VP60.

Active immunization with purified viral VP60 or the VP60 fusion protein (oral or parenteral vaccination) expressed in a baculovirus system has been shown to induce protection of rabbits against a lethal challenge with RHDV (Parra and Prieto 1990; Plana-Durán et al. 1996, respectively).

Plant-based production of VP60 was first achieved by expression using a plum pox potyvirus (PPV)-based vector (Fernández-Fernández et al. 2001). The foreign sequence was cloned between the NIb replicase and the coat protein cistrons of PPV, where posttranslational proteolytic cleavage releases the VP60 polypeptide during virus infection in the plant. Immunization of rabbits subcutaneously with leaf extracts (with an oily adjuvant) of *Nicotiana clevelandii* plants infected with plant virus chimera induced an effective immune response that protected animals against a lethal, intranasal challenge with RHDV.

The VP60 protein has also been produced in transgenic potato plants under control of the CaMV 35S promoter or a modified 35S promoter that included two copies of a strong transcriptional enhancer. Both promoters produced detectable levels of recombinant VP60, with higher levels being produced with the modified promoter (Castañón et al. 1999). Rabbits immunized parenterally with leaf extracts from plants carrying the modified promoter showed high anti-VP60 antibody titers and were fully protected against the hemorrhagic disease (Castañón et al. 2002).

Martín-Alonso et al. (2003) reported the development of transgenic potatoes producing up to 3.5 μg VP60/ mg of TSP in the tuber, levels of expression significantly higher than that of the TGEV glycoprotein S(N-gS) in transgenic potato tubers found by Gomez et al. (2000), but similar to levels reported by Mason et al. (1996) for transgenic potato tubers expressing Norwalk virus capsid protein. Oral immunization with potato tuber extracts (four doses of reconstituted lyophilized extracts in water; extracts containing either 100 or 500 μg of VP60) was performed using a syringe. Only two animals who received the 500-μg dose were seropositive for anti-VP60 antibodies after the third dose. Those receiving the 100-μg dose produced no detectable immune response, even though the equivalent amount of antigen produced an immune response when presented intramuscularly (Castañón et al. 1999, 2002). Rabbits were challenged with RHDV to evaluate the protective efficacy of the vaccination regime; only the rabbit having the highest anti-VP60 antibody titer survived, therefore achieving only a low level of protection. Several other animals that received the high (500 μg) dose of antigen did not survive the virus challenge; however, they survived longer than those not receiving the vaccine.

Additional Diagnostic Reagents and Vaccines for Viral Diseases

Group A Rotavirus

Group A rotavirus is one of the most important causes of severe viral diarrhea in humans and animals. It is a member of the *Rotavirus* genus of the family Reoviridae and is a multicomponent, double-stranded RNA virus that has wheel-like capsids in which spikes radiate from the inner capsid to the smooth viral outer capsid (Levy et al. 1994). The virions are not enveloped and three proteins make up the outer capsid, and four to six proteins, the core. VP6, the major capsid protein of the virus, is located on the inner capsid, and contains the common antigens of each rotavirus serogroup. VP6 has also been shown to be a protective antigen in a mouse infection model (Choi et al. 1999).

It has previously been shown that VP6 could be expressed from a Potato virus X-based vector in *N. benthamiana* plants (O'Brien et al. 2000). Matsumura et al. (2002) reported the first production of an immunogenic VP6 protein of bovine group A rotavirus in transgenic potato plants as a candidate diagnostic reagent for disease detection. The maximum level of antigen production was 0.1% TSP in leaf tissue. When potato tuber extracts containing recombinant VP6 were injected intraperitoneally into mice, anti-VP6 antibodies were detected in mouse serum. Sera were able to detect the purified GAR 22R strain in ELISA and in Western blots. In addition to its potential use as a diagnostic reagent to detect rotavirus serogroups, the VP6 antigen could also be developed into a subunit vaccine.

Wigdorovitz et al. (2004) reported on the development of an edible peptide vaccine for bovine rotavirus, but in this case an immunodominant peptide derived from the VP4 protein of bovine rotavirus (BRV) was expressed in transgenic alfalfa plants as a translational fusion with β-glucuronidase, which acts as a carrier protein. VP4 is an outer capsid protein forming spikes that emerge from the virion surface layer and is implicated in absorption of the virus to epithelial cells. Between 0.4 and 0.9 mg of fusion protein per gram of TSP of leaf extracts was produced. Mice were vaccinated intraperitoneally with crude leaf extracts or orally with freshly harvested transgenic leaves. Both sets of mice developed an immune response to BRV. Pups born to dams of both sets of mice showed a significant degree of protection when challenged with BRV.

Bovine Herpesvirus Type 1

Bovine herpesvirus type 1 (BHV-1) is the causative agent of a group of respiratory and reproductive disorders in cattle, commonly referred to as infectious bovine rhinotracheitis (Kahrs 1977). It affects adult and young animals and is common worldwide. The herpes virion consists of a dense core that is covered by an icosohedral capsid. The structure is also covered by a lipid bilayer envelope with short glycoprotein spikes on the surface (Levy et al. 1994). The genome is a double-stranded DNA that is linear but is in a circular form in the capsid. Several proteins are present on the surface of the virus particle, of which half are glycoproteins. Current vaccines for BHV-1 are formulated from inactivated or modified live virus and have the disadvantages of being poorly immunogenic or producing clinical disease if poorly inactivated, respectively. Alternative vaccination strategies using viral components, including glycoprotein D (gD) to induce protective immune responses, have been explored (van Drunen Littel-van den Hurk et al. 1993).

In a report by Pérez-Filgueira et al. (2003), the TMV viral-based expression system was used to produce a truncated, cytoplasmic subunit form of BHV-1 gD protein in plants. The amount of recombinant protein was estimated at 15–20 μg per gram of fresh leaf tissue. Crude extracts of *N. benthamiana* leaves inoculated with the recombinant virus were used to parenterally vaccinate mice (0.2 g per dose containing 2 μg of antigen) and cattle (5 g per dose). Both humoral and cellular-specific responses recognizing the gD antigen were induced, and the candidate vaccine was able to induce protection in the natural bovine host to challenge, intranasal viral infection with the BHV-1 LA strain. Protection in cattle was manifested by reduced amounts of excreted virus in nasal fluids, as compared to BHV-1 vaccinated cows, and later and milder clinical symptoms of the disease in vaccinated animals.

Rinderpest Virus

Rinderpest is an acute highly contagious, often fatal, disease of cattle and other domestic and wild ruminants. The disease affects the gastrointestinal and respiratory systems. It is caused by rinderpest virus (RPV), a member of the genus *Morbillivirus* of the family Paramyxoviridae. A highly effective live, attenuated virus vaccine is available and the disease has been successfully eradicated from most parts of the world (Plowright 1962). However, a few foci of the disease still exist in parts of Africa, the Middle East, and South Asia. The disease remains a threat to livestock in developing countries. The difficulty in maintaining a cold chain for the vaccine results in failure of vaccination in the hot regions where rinderpest is endemic. Attempts at making thermostable whole virus vaccines have been made using animal virus vectors (Romero et al. 1994).

The lipid envelope of RPV contains two glycoproteins, F (fusion, 65 kDa) and H (hemagglutinin, 80 kDa), which form spike-like structures on the surface of the particle. Virus infection is initiated by the action of these two proteins: H mediates the attachment of the virus to the host cell membrane and F mediates virus penetration into the host cell and virus-induced cell fusion and hemolysis. These proteins, which are known to be highly immunogenic, are important targets for the host cell response and confer protective immunity. Efforts have been made to develop subunit vaccines of recombinant vaccinia virus or capripox expressing the H and F proteins, and long-term immunity with these vaccines has been demonstrated (Yamanouchi et al. 1993; Romero et al. 1994, respectively). Although these vaccines are effective, their use may become prohibitively expensive due to the high cost of the cell culture used to produce the recombinant animal virus vaccines.

Khandelwal and colleagues (Khandelwal et al. 2003a, b, 2004; Satyavathi et al. 2003) have reported attempts to produce a less expensive, yet effective, vaccine by expression of the H protein in transgenic tobacco, peanut, and pigeon pea and oral delivery of the candidate vaccine in a mouse model and in cattle. The H protein expressed in tobacco was immunogenic and elicited a specific humoral response in an experimental mouse model (Khandelwal et al. 2003a).

Pigeon pea, also known as regram, is used as a food and fodder crop, with the foliage being used for animal feed after the seeds have been harvested for human consumption. Satyavathi et al. (2003) transformed pigeon pea with the H protein under transcriptional control of the CaMV 35S promoter. Levels of H protein in transgenic leaves reached 0.49% TSP. In peanuts transformed with the same construct, the expression level of H protein was in the range of 0.2%–1.3% TSP in leaf extracts (Khandelwal et al. 2003b). The peanut-derived H protein was immunologically active when delivered orally or parenterally in the absence of adjuvant in an experimental mouse model system (Khandelwal et al. 2004). When cattle were fed with transgenic peanut leaves at weekly intervals for 3 weeks with 5–7.5 g of leaf tissue, H-specific antibody was detected in serum of immunized cattle, and the serum neutralized RPV virus infectivity in vitro (Khandelwal et al. 2003b).

Classical Swine Fever Virus

Classical swine fever virus (CSFV), also known as hog cholera virus, is related to bovine viral diarrhea virus and belongs to the genus *Pestivirus* of the family Flaviviridae (Meyers and Thiel 1996). CSFV is a highly contagious viral disease of swine that occurs in an acute, subacute, chronic, or persistent form. In the acute form, the disease is characterized by high fever, severe depression, multiple superficial and internal hemorrhages, and high morbidity and mortality (Kleiboeker 2002). In the chronic form, the signs of depression, anorexia, and fever are less severe than in the acute form, and recovery is occasionally seen in mature animals. Transplacental infection with viral strains of low virulence often results in persistently infected piglets, which constitute a major cause of virus dissemination to noninfected farms.

CSFV is an enveloped RNA virus with a viral nucleocapsid containing a single molecule of genomic, positive-sense RNA complexed with a single polypeptide. The viral capsid is surrounded by a lipid bilayer that contains two proteins, one of which is the 51- to 60-kDa E2 protein, which determines the serological specificity of the virus. Legocki et al. (2005) sequenced the E2 gene from two strains of CSFV and constructed trangenic lettuce and alfalfa plants in which the gene was under the control of the CaMV35S promoter. One gram of lyophilized lettuce contained

10 μg of E2 antigen; when coupled to ubiquitin, the yield increased to 160 μg of antigen per gram of dry tissue. Preliminary testing of antigenicity by oral administration to mice and rats revealed an increase in IgG and IgA antibodies against the antigen after the second immunization.

Hantavirus

Hantaviruses (HV) are an emerging threat to animal and human health as new outbreaks of hantavirus infections are occurring with more frequency (Elliott et al. 1994). Hantaviruses cause the clinical syndromes in humans known as hemorrhagic fever with renal syndrome, and hantavirus pulmonary syndrome with high fatality worldwide. The individual hantavirus genotypes are carried by specific rodent hosts that do not exhibit clinical signs of infection. The major mode of transmission to is by aerosolized excreta of HV-infected rodents. Anyone who comes into contact with the virus is susceptible to the disease, and those especially at risk for infection include infants, the immunocompromised, farmers, veterinarians, rodent breeders, and zoo/wildlife/primate and other animal healthcare workers. There is no commercially available hantavirus vaccine; however, inactivated virus vaccines are currently being developed and tested (Cho et al. 2002; Hjelle 2002). Although cats and dogs are not known to be hosts for hantavirus, they may bring infected rodents into contact with humans and, therefore, development of oral, plant-based vaccine as bait in rodents would seem an effective strategy to control the disease in both rodents and humans.

Hantaviruses are negative-sense, single-stranded RNA, enveloped viruses that belong to the genus *Hantavirus* of the family Bunyaviridae. The three-segmented genome encodes RNA-dependent RNA polymerase on the L-RNA segment, two surface glycoproteins G1 and G2 on the M-RNA segment, and the viral nucleocapsid protein (N) on the S-RNA segment. Major antigenic domains are located on the N protein (Gött et al. 1997). The N protein and the G1 and G2 glycoproteins are promising candidates for development of subunit vaccines using plant-based expression strategies. Kehm et al. (2001) recently described the development of transgenic tobacco and potato plants expressing the Puumala virus N protein. The recombinant protein was expressed at 1 ng/4 μg of dried leaf and root tissues of transgenic potato. Rabbits immunized intraperitoneally with leaf extracts from tobacco and potato plants produced anti-N protein serum. Oral immunization of mice is under investigation. It has not yet been reported if the recombinant proteins are able to elicit effective protective immune responses in test animals.

Infectious Bronchitis Virus

Infectious bronchitis virus (IBV) is the pathogen causing chicken infectious bronchitis (IB), an acute, highly contagious respiratory disease of young chicks. The disease is controlled by serotype-specific vaccines. Outbreaks of IB occur due to the lack of cross-protection of the vaccines between serologically distinct viruses. Zhou et al. (2004) developed transgenic potato lines expressing the full-length spike protein of IBV. IBV, like TGEV above, is a member of the family *Coronoviridae*, order Nidovirales, and its genome encodes three major structural genes, one of which is the spike (S) protein. The entire 120-kDa (S) protein was expressed in potato tubers, and the plant-derived S protein retained reactivity to IBV antisera on Western blots. Following oral or intramuscular inoculations of chicks with potato tuber extracts containing the S protein (2.5 µg/g tuber tissue), detectable levels of serum neutralizing antibodies were seen. In addition, the chicks were protected against challenge with virulent IBV.

Vaccines and Therapeutics Against Bacterial and Parasitic Diseases

Swine Edema

Swine edema, also called gut edema, can cause morbidity and mortality in piglets. It is caused by certain serotypes of enterotossimic *Escherichia coli* (0138, 0139, 0141) that are able to produce a powerful, Shiga-like toxin. The toxin damages the walls of small blood vessels, including those of the brain, and causes fluid, or edema, to accumulate in tissues of the stomach and the large bowel. Damage to blood vessels in the brain results in characteristic nervous signs. Flexible filaments, known as F18 fimbriae, are expressed by various strains of *E. coli* causing disease in humans and animals, and have been reported to be an important virulence factor related to edema (Rippinger et al. 1995). The backbone of fimbriae consists of about 1,000 copies of a polypeptide that are arranged in an open helix. Colonization of pigs with the bacterial strains results in high levels of antifimbriae antibodies, especially IgA. As a result, there have been attempts made to vaccinate with the fimbrial proteins.

Rossi et al. (2003) reported the isolation of the F18 fimbriae gene from genomic DNA of *E. coli* isolated from pigs that died of edema. The gene was cloned into plant transformation vectors, and transgenic tobacco seeds expressing up to 0.1% TSP of F18 adhesive fimbriae were produced. No animal tests were conducted in this study. The tobacco seeds may offer a source of oral vaccine to protect against the disease.

Porcine Taenia Solium

Porcine *Taenia solium*, the pork tapeworm, is a long, flat, ribbon-like parasite that averages 6–10 feet long but can reach 30 feet in length. It causes a major parasitic disease known as taeniasis/cysticercosis that affects humans and pigs and is prevalent in poor sanitary conditions and rustic rearing of pigs. Cysticerci (one stage of the life cycle) may localize in the central nervous system of humans, causing neurocysticercosis, a major health problem in developing countries. Pigs play the role of obligatory intermediate host in the life cycle of the parasite; therefore, control of the disease in pigs by vaccination may reduce human infection. A number of vaccines have been investigated, including the use of native or recombinant antigens, or peptides, derived from different stages of the tapeworm life cycle (Plancarte et al. 1999). Three well-defined peptides of 18, 12, and 8 amino acids have been developed into a synthetic vaccine, SPvac, that is very effective under field conditions (Huerta et al. 2001), but is costly to produce. In a search for less costly alternatives, Sciutto et al. (2002) engineered the peptides to be expressed in recombinant filamentous phage (M13) and in transgenic carrot and papaya plants. Candidate antigens were identified from cDNA expression libraries of a related parasite, *T. crassiceps*, by screening with sera from *T. solium*-cysticerci-infected pigs. Three peptides were identified from three of the expression products as being antigenic, and were shown to give high levels of protection in a mouse model. These protective peptides are distributed among the different life stages of the *T. solium* parasite. No data are yet available on expression levels in plants or effectiveness of the plant-based vaccine.

Bovine Pneumonic Pasteurellosis

Bovine pneumonic pasteurellosis (PP), also known as shipping fever or bovine respiratory disease complex, is a major cause of sickness in transported cattle (Yates 1982). The bacterium, *Mannheima (Pasteurella) haemolytica* serotype A1 is the principal microorganism responsible for the disease. It inhabits the upper respiratory tract of healthy, unstressed calves. The disease occurs due to a combination of the presence of the bacteria, respiratory disease viruses, and stress. Traditional vaccination may involve subcutaneous injection of live virus, tissue culture-derived killed virus, orally administered bacterial culture supernatants, outer membrane protein preparations, and subunit vaccines (e.g., Sreevatsan et al. 1996). These vaccines provide various degrees of protection; however, they all involve herding and restraint of animals.

A noninvasive alternative vaccine candidate for PP has been developed by Lee et al. (2001) and is based on the expression of one of the major virulence factors of *M. haemolytica* A1. The leukotoxin (Lkt) is a protein that is secreted by the bacterium and acts as a pore-forming cytolysin that inserts into the plasma membrane of host cells, leading to cell lysis. Resulting tissue damage leads to pneumonia and death of animals. A recombinant derivative of Lkt, Lkt50, lacking the putative hydrophobic transmembrane domains, was engineered as an N-terminal fusion with modified green fluorescent protein (mGFP) and containing an N-terminal signal sequence and C-terminal endoplasmic reticulum retention signal. This chimera was used to generate transgenic white clover (*Trifolium repens* L.) expressing the chimeric protein from the CaMV 35S promoter at levels of 1% Lkt50-GFP (equivalent to 18 μg/g fresh weight tissue). Lkt-GFP-enriched fractions prepared from leaf extracts were able to induce an immune response to authentic Lkt in rabbits when delivered by intramuscular injection. The resulting antibodies were able to neutralize Lkt. According to the report, experiments are in progress to assess the immunogenicity of the candidate vaccine in cattle and to test the efficacy of feeding the transgenic clover to cattle in stimulating a mucosal immune response to Lkt.

Fasciola Hepatica

The liver fluke, *Fasciola hepatica*, is a common parasite of sheep and cattle, and also infects humans worldwide. Fascioliasis in sheep, cattle, and goats results in animals which show low productivity and higher mortality, causing severe economic losses. Among the proteins released by flukes are potent proteases required for parasite metabolism. Previous reports suggested that cysteine proteases could be used as protective antigens when injected intramuscularly into rats (Wedrychowicz et al. 2003). Legocki et al. (2005) engineered a DNA fragment encoding the catalytic domain of the cysteine protease of *F. hepatica* into plant transformation vectors and used to transform lettuce. When fused to a ubiquitin protein sequence, antigen expression was 100 μg of catalytic domain per gram of dried tissue. Both IgG and IgA antibodies were present in serum of mice that had been fed the transgenic plant material.

Coliform Mastitis

Coliform mastitis is one of the most common forms of environmental mastitis in dairy cows, accounting for 40%–50% of all clinical cases of mastitis (Hogan and Larry Smith 2003). Significant losses also occur in goats, sheep, and pigs. Important Gram-negative organisms in mastitis include *E. coli*, *Klebsiella pneumonia*, *Serratia marcescens*, and *Pseudomonas aeruginosa*. *E. coli* is shed from the intestinal tract. All of the Gram-negative organisms can also be found in the environment such as in soil, water, and bedding, and are common in dairy operations. If left unchecked, coliform mastitis can cause severe economic losses from a number of factors: reduced milk production and quality, increased labor and treatment costs, and increased culling rate and death losses. In the absence of effective vaccines, current control strategies rely heavily upon antibiotics and topical germicidal chemicals.

The bovine CD14 antigen is a high-affinity receptor for the complex of lipopolysaccharide (LPS, endotoxin) and LPS-binding protein. The secreted form of CD14 (sCD14) binds and neutralizes LPS from *E. coli* and other coliform bacteria and prevents development of acute endotoxin shock in cows as well as intramammary infection by coliform organisms (Lee et al. 2003). A CD14 recombinant gene was incorporated into a plant virus vector for transient expression in *N. benthamiana* (Nemchinov et al. 2006). Western blots probed with CD14-specific antibodies demonstrated that crude plant extracts as well as affinity-purified samples contained immunoreactive recombinant protein of predicted molecular mass. Biological activity of the plant-derived sCD14 was demonstrated in vitro by induction of apoptosis and interleukin-8 production in bovine endothelial cells and in vivo in bovine udders as shown by an increased leukocyte response in the presence of LPS.

Immunocontraceptive Vaccines and Follicle Stimulating Hormone

Fertility control by vaccination is a management practice that has been used on captive wildlife species, such as feral, or wild, horses (Kirkpatrick et al. 1992). The product in this case was injected whole porcine zona pellucidae (ZP), a thick extracellular matrix surrounding the mammalian ovum (egg) which binds sperm, harvested from pigs at slaughter (Barber and Fayrer-Hosken 2000). Antibodies are induced against reproductive self-antigens present on the ZP, resulting in a reduction in fertility.

Smith et al. (1997) proposed the use of oral, plant-derived immunocontraceptive vaccines for management of free-ranging wildlife species. In New Zealand, the introduced brushtail possum (*Trichosurus vulpecula*) causes major economic and environmental damage. In a recent report by Polkinghorne et al. (2005), the feasibility of using a plant-based immunocontraceptive composed of recombinant antigens derived from the ZP for control of possums was investigated. Females injected with porcine ZP showed reduced fertility and production of anti-ZP antibodies. Recombinant antigens (glycoproteins ZP2 and ZP3) from the ZP were cloned and expressed in bacteria; females injected with the purified proteins showed 75%–80% reduction in fertility. Although the recombinant ZP antigens have not yet been produced in a plant, female possums fed with transgenic potato tubers expressing the model antigen LT-B (heat-labile *E. coli* enterotoxin) (Mason et al. 1998) expressed systemic antibodies and antibody-secreting cells against LT-B. Future research includes the expression of the ZP antigens in trangenic carrot root and delivery of the vaccines in an oral bait.

At the other end of the spectrum, superovulation and embryo transfer is widely used to improve success of reproduction in economically important animals, including cattle. Superovulation is usually induced using pregnant mare serum gonadotropin or pituitary-derived follicle stimulating hormone (FSH), both of which have the disadvantage of potentially infectious agents in the purified preparations. Dirnberger et al. (2001) used TMV transient expression in *N. benthamiana* to produce the single-chain version of bovine FSH, a protein that requires extensive N-glycosylation for proper folding, activity, and stability. The protein was secreted to the extracellular compartment and up to 3% of TSP was produced. Although the recombinant protein contained plant glycans, it retained significant bioactivity in mice, though much lower than that of pregnant mare serum gonadotropin.

Regulatory Issues and Future Prospects

The regulatory considerations for products made in bioengineered plants, using either engineered viruses or transgenic plants, are, for the most part, the same as those for other therapeutics or vaccines, except for issues that may be unique for production of the products in plants (Stein and Webber 2001; Peterson and Arntzen 2004). Compliance of plant-based vaccines to Good Manufacturing Practices (GMP), and control of toxicity, dose, lot-to-lot consistency, possible allergic responses and immune tolerance are all to be considered for testing and commercialization of plant-based biologics. Draft guidance for drugs, biologics, and medical devices derived from engineered plants has been issued by the United States Food and Drug Administration Center for Veterinary Medicine and is currently in the comment phase (http://www.fda.gov/cvm/biotechnology/bio_feeds.html). Other issues that may need to be addressed relate to the use of bioengineered feed in animals where meat or milk is destined to be used as human food.

Future directions for veterinary vaccines include the development of oral, multivalent vaccines to replace injectable vaccine mixes. As an example, multivalent vaccines for poultry viruses, expanded vaccine targets to reduce human exposure to the pathogens that may be in present in meat products (e.g., *E. coli* 0157:H7 and *Salmonella*) or for diseases that are currently uncontrollable by vaccination. Oral neutralizing antibodies, similar to those produced as a prophylaxis against rabies in humans (Ko et al. 2003), are also potential therapeutic reagents for animal disease control. Among other potential vaccines that are currently being investigated, but not yet reported in the literature, are oral vaccines for farmed fish, and vaccines for equine herpesvirus.

The potential for the use of plant-derived biologics in animal health is high. However, in spite of the potential of plants as bioreactors, one would also need to show that the plant-derived product is clearly advantageous either for efficacy, ease of delivery and/or cost. Several of the reports reviewed in this chapter provide the proof of principle that the products are efficacious. However, cost is clearly a factor when replacements are considered for traditional veterinary vaccines, some of which are currently available for pennies per dose. With that said, the most likely near-term possibilities for commercialization of plant-derived vaccines and therapeutics will likely be veterinary products.

References

Barber MR, Fayrer-Hosken RA (2000) Possible mechanisms of mammalian contraception. J Reprod Immunol 46:103–124

Berinstein A, Tami C, Taboga O, Smitsaart E, Carrillo E (2000) Protective immunity against foot-and-mouth disease virus induced by a recombinant vaccinia virus. Vaccine 18:2231–2238

Bowersock TL, Martin S (1999) Vaccine delivery to animals. Adv Drug Deliv Rev 38:167–194

Brochier B, Thomas I, Baiduin B, Leveau T, Pastoret PP, Languet B, Chappuis G, Desmettre P, Balncou J, Artois M (1990) Use of a vaccinia-rabies recombinant virus for the oral vaccination of foxes against rabies. Vaccine 8:101–104

Brown F (1992) Vaccination against foot and mouth disease virus. Vaccine 10:1022–1026

Carrillo C, Wigdorovitz A, Oliveros JC, Zamorano PI, Sadir AM, Gómez N, Salinas J, Escribano JM, Borca MV (1998) Protective immune response to foot-and-mouth disease virus with VP1 expressed in transgenic plants. J Virol 72:1688–1690

Carrillo C, Wigdorovitz A, Trono K, Dus Santos MJ, Castañón S, Sadir AM, Ordas R, Escribano JM, Borca MV (2001) Induction of a virus-specific antibody response to foot and mouth disease virus using the structural protein VP1 expressed in transgenic potato plants. Viral Immunol 14:49–57

Casal JI, Langeveld JPM, Cortes E, Schaaper WWM, van DIJK E, Vela C, Kamstrup S, Meloen RH (1995) Peptide vaccine against canine parvovirus: identification of two neutralization subsites in the N terminus of VP2 and optimization of the amino acid sequence. J Virol 69:7272–7277

Castañón S, Marin MS, Martin-Alonso JM, Boga R, Casais R, Humara JM, Ordas RJ, Parra F (1999) Immunization with potato plants expressing VP60 protein protects against rabbit hemorrhagic disease virus. J Virol 73:4452–4455

Castañón S, Martin-Alonso JM, Marin MS, Boga JA, Alonso P, Parra F, Ordas RJ (2002) The effect of the promoter on expression of VP60 gene from rabbit hemorrhagic disease virus in potato plants. Plant Sci 162:87–95

Cho HW, Howard CR, Lee HW (2002) Review of inactivated vaccine against hantaviruses. Intervirology 45:328–333

Choi AH, Basu M, McNeal MM, Clements JD, Ward RL (1999) Antibody-independent protection against rotavirus infection of mice stimulated by intranasal immunization with chimeric VP4 or VP6 protein. J Virol 73:7574–7581

Dalsgaard K, Uttenthal A, Jones TD, Xu F, Merryweather A, Hamilton WDO, Langeveld JPM, Boshuizen RS, Kamstrup S, Lomonossoff GP, Porta C, Vela C, Casal JI, Meloen RH, Rodgers PB (1997) Plant-derived vaccine protects target animals against a viral disease. Nat Biotechnol 15:248–252

Dirnberger D, Steinkellner H, Abdennebi L, Remy JJ, van de Wiel D (2001) Secretion of biologically active glycoforms of bovine follicle stimulating hormone in plants. Eur J Biochem 268:4570–4579

Dus Santos MJ, Wigdorovitz A, Trono K, Rios RD, Franzone PM, Gil F, Moreno J, Carrillo C, Escribano JM, Borca MV (2002) A novel methodology to develop a foot and mouth disease virus (FMDV) peptide-based vaccine in transgenic plants. Vaccine 20:1141–1147

Dus Santos MJ, Carrillo C, Ardila F, Ríos RD, Franzone P, Piccone ME, Wigdorovitz A, Borca MV (2005) Development of transgenic alfalfa plants containing the foot and mouth disease virus structural polyprotein gene P1 and its utilization as an experimental immunogen. Vaccine 23:1838–1843

Elliott LH, Ksiazek TG, Rollin PE, Spiropoulou CF, Morzunov S, Monroe M, Goldsmith CS, Humphrey CD, Zaki SR, Krebs JW, et al (1994) Isolation of the causative agent of hantavirus pulmonary syndrome. Am J Trop Med Hyg 51:102–108

Fernández-Fernández MR, Martinez-Torrecuadrada JL, Casal JI, Garcia JA (1998) Development of an antigen presentation system based on plum pox potyvirus. FEBS Lett 427:229–235

Fernández-Fernández MR, Mouriño M, Rivera J, Rodríguez F, Plana-Durán J, García JA (2001) Protection of rabbits against rabbit hemorrhagic disease virus by immunization with the VP60 protein expressed in plants with a potyvirus-based vector. Virology 280:283–291

Gay CG, Salt J, Balaski C (2003) Challenges and opportunities in developing and marketing vaccines for OIE List A and emerging animal diseases. Dev Biol (Basel) 114:243–250

Gil F, Brun A, Wigdorovitz A, Catala R, Martinez-Torrecuadrada JL, Casal I, Salinas J, Borca MV, Escribano JM (2001) High-yield expression of a viral peptide vaccine in transgenic plants. FEBS Lett 488:13–17

Gött P, Zöller L, Yang S, Darai G, Bautz EKF (1997) A major antigenic domain of hantaviruses is located on the aminoproximal site of the viral nucleocapsid protein. Virus Genes 14:31–40

Gómez N, Carillo C, Salinas J, Parra F, Borca MV, Escribano JM (1998) Expression of immunogenic glycoprotein S polypeptides from transmissible gastroenteritis coronavirus in transgenic plants. Virology 249:352–358

Gómez N, Wigdorovitz A, Castañón S, Gil F, Ordás R, Borca MV, Escribano JM (2000) Oral immunogenicity of the plant derived spike protein from swine-transmissible gastroenteritis coronavirus. Arch Virol 145:1725–1732

Hjelle B (2002) Vaccines against hantaviruses. Expert Rev Vaccines 1:373–384

Hogan J, Larry Smith K (2003) Coliform mastitis. Vet Res 34:507–519

Huerta M, Aluja A, Fragaso G, Toledo A, Villalobos N, Hernández M, Gevorkian F, Acero G, Díaz A, Alvarez I, Avila R, Beltrán C, García G, Martínez JJ, Larralde C, Sciutto E (2001) Synthetic peptide vaccine against *Taenia solium* pig cysticercosis: successful vaccination in a controlled field trial in rural Mexico. Vaccine 20:262–266

Kahrs RF (1977) Infectious bovine rhinotracheitis: a review and update. J Am Vet Med Assoc 171:1055–1064

Kehm R, Jakob NJ, Welzel TM, Tobiasch E, Viczian O, Jock S, Geider K, Süle S, Darai G (2001) Expression of immunogenic Puumala virus nucleocapsid protein in transgenic tobacco and potato plants. Virus Genes 22:73–83

Khandelwal A, Lakshmi Sita G, Shaila MS (2003a) Expression of hemagglutinin protein of rinderpest virus in transgenic tobacco and immunogenicity of plant-derived protein in a mouse model. Virology 308:207–215

Khandelwal A, Sita GL, Shaila MS (2003b) Oral immunization of cattle with hemagglutinin protein of rinderpest virus expressed in transgenic peanut induces specific immune responses. Vaccine 21:3282–3289

Khandelwal A, Renukaradhya GJ, Rajasekhar M, Sita GL, Shaila MS (2004) Systemic and oral immunogenicity of hemagglutinin protein of rinderpest virus expressed by transgenic peanut plants in a mouse model. Virology 323:284–291

Kirkpatrick JF, Liu IMK, Truner JW, Naugle R, Keiper R (1992) Long-term effects of porcine zona pellucidae immunocontraception on ovarian function in feral horses (*Equus caballus*). J Reprod Fertil 94:437–444

Kleiboeker SB (2002) Swine fever: classical swine fever and African swine fever. Vet Clin North Am Food Anim Pract 18:431–451

Ko K, Tekoah Y, Rudd PM, Harvey DJ, Dwek RA, Spitsin S, Hanlon CA, Rupprecht C, Dietzschold B, Golovkin M, Koprowski H (2003) Function and glycosylation of plant-derived antiviral monoclonal antibody. Proc Natl Acad Sci U S A 100:8013–8018

Lamphear BJ, Streatfield SJ, Jilka JM, Brooks CA, Barker DK, Turner DD, Delaney DE, Garcia M, Wiggins B, Woodard SL, Hood EE, Tizard IR, Lawhorn B, Howard JA (2002) Delivery of subunit vaccines in maize seed. J Control Release 85:169–180

Lamphear BJ, Jilka JM, Kesl L, Welter M, Howard JA, Streatfield SJ (2004) A corn-based delivery system for animal vaccines: an oral transmissible gastroenteritis virus vaccine boosts lactogenic immunity in swine. Virology 22:2420–2424

Langeveld JP, Casal JI, Osterhaus AD, Cortes E, de Swart R, Vela C, Dalsgaard K, Puijk WC, Schaaper WM, Meloen RH (1994) First peptide vaccine providing protection against viral infection in the target animal: studies of canine parvovirus in dogs. J Virol 68:4506–4513

Langeveld JPM, Kamstrup S, Uttenthal A, Strandbygaard B, Vela C, Dalsgaard K, Beekman NJ, Meloen RH, Casal JI (1995) Full protection in mink against mink enteritis virus with new generation canine parvovirus vaccines based on synthetic peptide or recombinant protein. Vaccine 13:1033–1037

Langeveld JPM, Brennan FR, Martinez-Torrecuadrada JL, Jones TD, Boshuizen RS, Vela C, Casal JI, Kamstrup S, Dalsgaard K, Meleon RH, Bendig MM, Hamilton WDO (2001) Inactivated recombinant plant virus protects dogs from a lethal challenge with canine parvovirus. Vaccine 19:3661–3670

Laude H, Rasschaert D, Delmas B, Godet M, Gelfi J, Charley B (1990) Molecular biology of transmissible gastroenteritis virus. Vet Microbiol 23:147–154

Lee JW, Paape MJ, Elsasser TH, Zhao X (2003) Recombinant soluble CD14 reduces severity of intramammary infection by *Escherichia coli*. Infect Immun 71:4034–4039

Lee RWH, Strommer J, Hodgins D, Shewen PE, Niu Y, Lo RYC (2001) Towards development of an edible vaccine against bovine pneumonic pasteruellosis using transgenic white clover expressing a *Mannhemia haemolytica* AS leukotoxin 50 fusion protein. Infect Immun 69:5786–5793

Legocki AB, Miedzinska K, Czaplińska M, Plucieniczak A, Wędrychowicz H (2005) Immunoprotective properties of transgenic plants expressing E2 glycoprotein from CSFV and cysteine protease from *Fasciola hepatica*. Vaccine 23:1844–1846

Levy JA, Fraenkel-Conrat H, Owens RA (eds) (1994) Virology, 3rd edn. Prentice-Hall, Englewood

Martin-Alonso JM, Castanon S, Alonso P, Parra F, Ordas R (2003) Oral immunization using tuber extracts from transgenic potato plants expressing rabbit hemorrhagic disease virus capsid protein. Transgenic Res 12:127–130

Mason HS, Ball JM, Shi JJ, Jiang X, Estes MK, Arntzen CJ (1996) Expression of Norwalk virus capsid protein in transgenic tobacco and potato and its oral immunogenicity in mice. Proc Natl Acad Sci U S A 93:5335–5340

Mason HS, Haq TA, Clements JD, Arntzen CJ (1998) Edible vaccine protects mice against *Escherichia coli* heat-labile enterotoxin (LT): potatoes expressing a synthetic LT-B gene. Vaccine 16:1336–1343

Mason PW, Chinsangaram J, Moraes MP, Mayr GA, Grubman MJ (2003) Engineering better vaccines for foot-and-mouth disease. Dev Biol (Basel). Transgenic Res 114:79–88

Matsumura T, Itchoda N, Tsunemitsu H (2002) Production of immunogenic VP6 protein of bovine group A rotavirus in transgenic potato plants. Arch Virol 147:1263–1270

McGarvey PB, Hammond J, Dinelt MM, Hooper DC, FU ZF, Dietzschold B, Koprowski H, Michaels FH (1995) Expression of the rabies virus glycoprotein in transgenic tomatoes. Biotechnology 13:1484–1487

Meloen RH, Hamilton WDO, Casal JI, Dalsgaard K, Langeveld JPM (1998) Edible vaccines. Vet Q 20 [Suppl] 3:92–95

Meyers G, Thiel HJ (1996) Molecular characterization of pestiviruses. Adv Virus Res 47:53–118

Modelska A, Dietzschold N, Sleysh N, Fu ZF, Steplewski K, Hooper DC, Koprowski H, Yusibov V (1998) Immunization against rabies with plant-derived antigen. Proc Natl Acad Sci U S A 95:2481–2485

Nemchinov LG, Paape M, Sohn EJ, Bannerman D, Zarlenga D, Hammond RW (2006) Bovine CD14 receptor produced in plants reduces severity of intramammary bacterial infection. FASEB J 20:1345–1351

Nicholas BL, Brennan FR, Martinez-Torrecuadrada JL, Casal JI, Hamilton WD, Wakelin D (2002) Characterization of the immune response to canine parvovirus induced by vaccination with chimaeric plant viruses. Vaccine 20:2727–2734

Nicholas BL, Brennan FR, Hamilton WD, Wakelin D (2003) Effect of priming/booster immunisation protocols on immune response to canine parvovirus peptide induced by vaccination with a chimaeric plant virus construct. Adv Virus Res 21:2441–2447

O'Brien GJ, Bryand CJ, Voogd C, Greenberg HB, Gardner RC, Ballamy AR (2000) Rotavirus VP6 expressed by PVX vectors in *Nicotiana benthamiana* coats PVX rods and also assembles into viruslike particles. Virology 270:444–453

Parra F, Prieto M (1990) Purification and characterization of a calicivirus as the causative agent of a lethal hemorrhagic disease in rabbits. J Virol 64:4013–4015

Peeters JE, Vandergheynst D, Geeroms R (1992) Vaccination contre la maladie hémorragique virale du lapin (VHD): effet protecteur de deux vaccins commerciaux. Cuni-Sciences 7:101–106

Pérez-Filgueria DMP, Zamorano PI, Domínguez MG, Taboga O, Del Médico Zajac MP, Puntel M, Romera SA, Morris TJ, Borca MV, Sadir AM (2003) Bovine herpes virus gD protein produced in plants using a recombinant tobacco mosaic virus (TMV) vectors possesses authentic immunogenicity. Vaccine 21:4201–4209

Peterson RKD, Arntzen CJ (2004) On risk and plant-based pharmaceuticals. Trends Biotechnol 22:64–66

Plana-Duran J, Bastons M, Rodriguez MJ, Climent I, Cortes E, Vela C, Casal I (1996) Oral immunization of rabbits with VP60 particles confers protection against rabbit hemorrhagic disease. Arch Virol 141:1423–1436

Plancarte A, Flisser A, Gauci CG, Lightowlers MW (1999) Vaccination against *Taenia solium* cysticercosis in pigs using native and recombinant oncosphere antigens. Int J Parasitol 29:643–647

Plowright W (1962) The application of monolayer tissue culture techniques in rinderpest research. II. The use of attenuated cultural virus as a vaccine for cattle. Bull Off Int Epizoot 57:253–277

Polkinghorne I, Hamerli D, Cowan P, Duckworth J (2005) Plant based immnocontraceptive control of wildlife-"potentials, limitations and possums". Vaccine 23:1847–1850

Porta C, Lomonossoff GP (1998) Scope for using plant viruses to present epitopes from animal pathogens. Rev Med Virol 8:25–41

Porta C, Spall VE, Findlay KC, Gergerich RC, Farrance CE, Lomonossoff GP (2003) Cowpea mosaic virus-based chimeras. Effects of inserted peptides on the phenotype, host range, and transmissibility of the modified viruses. Virology 310:50–63

Rippinger P, Bertschinger HU, Imberchts H, Nagy B, Sorg I, Stamm M, Wild P, Witting W (1995) Designations of F18ab and F18ac for related fimbrial types F107 2134P, and 8813 of *Escherichia coli* isolated from porcine post-weaning diarrhoea and from oedema disease. Vet Microbiol 45:281–295

Romero CH, Barrett T, Chamberlain RW, Kitching RP, Fleming MDN, Black DN (1994) Recombinant Capripoxvirus expressing the hemagglutinin protein of rinderpest virus: protection of cattle against rinderpest and lumpy skin disease viruses. Virology 204:425–429

Rossi L, Baldi A, Dell'Orto V, Fogher C (2003) Antigenic recombinant proteins expressed in tobacco seeds as a model for edible vaccines against swine oedema. Vet Res Commun 27 [Suppl 1]:659–661

Rupprecht C, Wiktor T, Johnston D, Hamir A, Dietzschold B, Wunner WH, Glickman LT, Koprowski H (1986) Oral immunization and protection of raccoons (*Procyon lotor*) with a vaccinia-rabies glycoprotein virus vaccine. Proc Natl Acad Sci U S A 83:7947–7950

Saif LJ, van Cott JL, Brim TA (1994) Immunity to transmissible gastroenteritis virus and porcine respiratory coronavirus infection in swine. Vet Immunol Immunopathol 43:89–97

Satyavathi VV, Prasad V, khandelwal A, Shaila MS, Sita GL (2003) Expression of the hemagglutinin protein of Rinderpest virus in transgenic pigeon pea [*Cajanus cajan* (L.) Millsp.] plants. Plant Cell Rep 21:651–658

Sciutto E, Fragoso G, Manoutcharian K, Gevorkian G, Rosas-Salgado G, Hernandez-Gonzalez M, Herrera-Estrella L, Cabrera-Ponce JL, Lopez-Casillas F, Gonzalez-Bonilla C, Santiago-Machuca A, Ruiz-Perez F, Sanchez J, Goldbaum F, Aluja A, Larralde C (2002) New approaches to improve a peptide vaccine against porcine *Taenia solium* cysticercosis. Arch Med Res 33:371–378

Smith G, Walmsley A, Polkinghorne I (1997) Plant-derived immunocontraceptive vaccines. Reprod Fertil Dev 9:85–89

Sreevatsan S, Ames TR, Werdin RE, Han SY, Maheswaran SK (1996) Evaluation of three experimental subunit vaccines against pneumonic pasteurellosis in cattle. Vaccine 14:147–154

Stein KE, Webber KO (2001) The regulation of biologic products derived from bioengineered plants. Curr Opin Biotechnol 12:308–311

Streatfield SJ, Howard JA (2003a) Plant-based vaccines. Int J Parasitol 33:479–493

Streatfield SJ, Howard JA (2003b) Plant production systems for vaccines. Expert Rev Vaccines 2:763–775

Streatfield SJ, Jilka JM, Hood EE, Turner DD, Bailey MR, Mayor JM, Woodard SL, Beifuss KK, Horn ME, Delaney DE, Tizard IR, Howard JA (2001) Plant-based vaccines: unique advantages. Vaccine 19:2742–2748

Streatfield SJ, Mayor JM, Barker DK, Brooks C, Lamphear BJ, Woodard SL, Beifuss KK, Vicuna DV, Massey LA, Horn ME, Delaney DE, Nikolov ZL, Hood EE, Jilka JM, Howard JA (2002)

Development of edible subunit vaccine in corn against enterotoxigenic strains of Escherichia coli. In Vitro Cell Dev Biol Plant 28:11–17

Sun M, Qian K, Su N, Chang H, Liu J, Chen G (2003) Foot-and-mouth disease virus VP1 protein fused with cholera toxin B subunit expressed in *Chlamydomonas reinhardtii* chloroplast. Biotechnol Lett 25:1087–1092

Torres JM, Alonso C, Ortega A, Mittal S, Graham F, Enjuanes L (1996) Tropism of human adenovirus type 5-based vectors in swine and their ability to protect against transmissible gastroenteritis coronavirus. J Virol 70:3770–3780

Tuboly T, Yu W, Bailey A, Degrandis S, Du S, Erickson L, Nagy E (2000) Immunogenicity of porcine transmissible gastroenteritis virus spike protein expressed in plants. Vaccine 18:2023–2028

Usha R, Rohll JB, Spall VE, Shanks M, Maule AJ, Johnson JE, Lomonossoff GP (1993) Expression of an animal virus antigenic site on the surface of a plant virus particle. Virology 197:366–374

van Drunen Littel-van den Hurk S, Parker MD, Massie B, van den Hurk JV, Harland R, Babiuk LA (1993) Protection of cattle from BHV-1 infection by immunization with recombinant glycoprotein gIV. Vaccine 11:25–35

Van Rensburg HG, Mason PW (2002) Construction and evaluation of a recombinant foot-and-mouth disease virus. Implications for vaccine production. Ann N Y Acad Sci 969:83–87

Walmsley AM, Kirk DD, Mason HS (2003) Passive immunization of mice pups through oral immunization of dams with a plant-derived vaccine. Immunol Lett 86:71–76

Wedrychowicz H, Lamparska M, Kesik M, Kotomski G, Mieszczanek J, Jedlina-Panasiuk L, Plucienniczak A (2003) The immune response of rats to vaccination with the cDNA or protein forms of the cysteine proteinase of *Fasciola hepatica*. Vet Immunol Immunopathol 94:83–93

Wigdorovitz A, Carrillo C, Dus Santos MJ, Trono K, Peralta A, Gómez MC, Rios RD, Franzone PM, Sadir AM, Escribano JM, Borca MV (1999a) Induction of a protective antibody response to foot and mouth disease virus in mice following oral or parental immunization with alfalfa transgenic plants expressing the viral structural protein VP1. Virology 255:347–353

Wigdorovitz A, Pérez Filgueria DM, Robertson N, Carrillo C, Sadir AM, Morris TJ, Borca MV (1999b) Protection of mice against challenge with foot and mouth disease virus (FMDV) by immunization with foliar extracts from plants infected with recombinant tobacco mosaic virus expressing the FMDV structural protein VP1. Virology 264:85–91

Wigdorovitz A, Mozgovoj M, Dus Santos MJ, Parreño V, Gómez C, Pérez-Filgueira DM, Trono KG, Ríios RD, Franzone PM, Fernández F, Carillo C, Babiuk LA, Escribano JM, Borca MV (2004) Protective lactogenic immunity conferred by an edible peptide vaccine to bovine rotavirus produced in transgenic plants. J Gen Virol 85:1825–1832

Wu L, Jiang L, Zhou Z, Fan J, Zhang Q, Zhu H, Han Q, Xu Z (2003) Expression of foot-and-mouth disease virus epitopes in tobacco by a tobacco mosaic virus-based vector. Vaccine 21:4390–4398

Yamanouchi K, Inui K, Sugimoto M, Asano K, Nishimaki F, Kitching RP, Takamatus H, Barrett T (1993) Immunisation of cattle with a recombinant vaccinia vector expressing the haemagglutinin gene of rinderpest virus. Vet Rec 132:152–156

Yates WDG (1982) A review of infectious bovine rhinotracheitis, shipping fever pneumonia and viral-bacterial synergism in respiratory disease of cattle. Can J Comp Med 46:225–263

Yusibov V, Modelska A, Steplewski K, Agadjanyan M, Weiner D, Hooper DC, Koprowski H (1997) Antigens produced in plants by infection with chimeric plant viruses immunize against rabies virus and HIV-1. Proc Natl Acad Sci U S A 94:5784–5788

Yusibov V, Hooper DC, Spitsin SV, Fleysh N, Kean RB, Mikheeva T, Deka D, Karasev A, Cox S, Randall J, Koprowski H (2002) Expression in plants and immunogenicity of plant virus-based experimental rabies vaccine. Vaccine 20:3155–3164

Zhou JY, Cheng LQ, Zhang XJ, Wu JX, Shang SB, Wang JY, Chen JG (2004) Generation of the transgenic potato expressing full-length spike protein of infectious bronchitis virus. J Biotechnol 111:121–130

Plant-Based Oral Vaccines: Results of Human Trials

C.O. Tacket

Contents

Abstract Vaccines consisting of transgenic plant-derived antigens offer a new strategy for development of safe, inexpensive vaccines. The vaccine antigens can be eaten with the edible part of the plant or purified from plant material. In phase 1 clinical studies of prototype potato- and corn-based vaccines, these vaccines have been safe and immunogenic without the need for a buffer or vehicle other than the plant cell. Transgenic plant technology is attractive for vaccine development because these vaccines are needle-less, stable, and easy to administer. This chapter examines some early human studies of oral transgenic plant-derived vaccines against enterotoxigenic *Escherichia coli* infection, norovirus, and hepatitis B.

C.O. Tacket(✉)
Center for Vaccine Development, University of Maryland School of Medicine,
685 West Baltimore St., Baltimore, MD 21201, USA
e-mail: ctacket@medicine.umaryland.edu

A.V. Karasev (ed.) *Plant-produced Microbial Vaccines.*
Current Topics in Microbiology and Immunology 332

Introduction

As new threats to public health emerge, the demand for effective, inexpensive, easy-to-administer, and most important, safe, vaccines will increase. In recent years, these new threats have included the agents of bioterror as well as new pathogens, such as SARS and H5N1 influenza.

The gastrointestinal mucosa is the largest site for induction of immune responses and use of this site to elicit protective immunity by oral immunization began with the oral polio vaccine in the 1960s. It makes sense to immunize the mucosa against pathogens that initiate their pathogenic processes at the mucosal surface. Some mucosally active vaccines have become successful public health tools, including oral polio vaccine, Ty21a oral typhoid vaccine (Vivotif), killed whole-cell/B subunit cholera vaccine (Dukoral), live attenuated cholera vaccine CVD 103-HgR (Orochol), and, most recently, intranasal cold-adapted influenza vaccine (FluMist).

Each of the successful mucosal vaccines, except the killed cholera vaccine, is a live attenuated version of the pathogen of interest. To construct the vaccine, defined and undefined virulence factors were purposely removed from the living pathogen. The complementary strategy is to identify one or a small number of antigens (subunits) of the pathogen that by themselves stimulate protective immune responses. This strategy has been hampered by the difficulty in identifying such protective antigens; however, the hepatitis B vaccine, consisting of hepatitis B surface antigen, is an excellent example of a highly successful parenteral subunit vaccine. The development of orally administered subunit vaccines has been hindered by the harsh environment of the human stomach and intestine. Several methods of protecting oral antigens have been developed and some tested in humans. These include encapsulation in microspheres, liposomes, ISCOMS, proteosomes, cochleates, and virosomes (Edelman 1997) or expression of the gene for the protective antigen in a commensal organism or an attenuated enteric pathogen (Grangette et al. 2002; Scheppler et al. 2002; Zegers et al. 1999; Pouwels et al. 1996; Devico et al. 2002; Orr et al. 1999; DiPetrillo et al. 1999; Tacket et al. 1997, 2000a; Nardelli-Haefliger et al. 1996).

One of the most promising novel techniques for production and oral delivery of subunit vaccines is by use of transgenic plants. Genes encoding antigens of interest from viral, bacterial, and parasitic pathogens can be expressed in the plant tissues, including the edible parts. Table 1 lists many of the vaccine antigens relevant to human disease that have been expressed in transgenic plants. The edible part of the transgenic plant can be ingested or the transgenic plant can be used as a bioreactor for large-scale, high-yield production of purified protein for oral or parenteral use (Giddings 2001; Giddings et al. 2000; Larrick and Thomas 2001). Examples of this application include production of monoclonal antibodies to provide passive immunotherapy (Ma et al. 1995, 1998; Chargelegue et al. 2000; Fischer et al. 2000; Zeitlin et al. 1998); an immunocontraceptive epitope (Walmsley et al. 2003; Smith et al. 1997), drugs (Giddings et al. 2000; Hood et al. 1997, 1999; Kusnadi et al. 1998a, 1998b; Daniell et al. 2001b), and autoantigens to induce oral tolerance in autoimmune disease (e.g., multiple sclerosis and type I diabetes) (Ma et al. 1997).

Table 1 Examples of vaccine antigens from human pathogens expressed in transgenic plants. Asterisk indicates phase 1 study has been done

Antigen	Plant	Reference
*Hepatitis B surface and core antigens	Tobacco; potato; cherry tomatillo; soybean; lettuce	Mason et al. 1992; Thanavala et al. 1995; Richter et al. 2000; Kapusta et al. 1999, 2001; Gao et al. 2003; Smith et al. 2002
Hepatitis E virus ORF2	Tomato	Ma et al. 2003
*Norwalk virus capsid protein	Potato	Mason et al. 1996; Tacket et al. 2000b
Rabies virus glycoprotein	Tomatoes; spinach	McGarvey et al. 1995; Modelska et al. 1998; Yusibov et al. 2002
Cytomegalovirus glycoprotein B	Tobacco	Tackaberry et al. 1999, 2003
HIV p24	Tobacco	Zhang et al. 2002
Measles virus hemagglutinin	Carrot; tobacco	Marquet-Blouin et al. 2003; Webster et al. 2002a, b
Human papillomavirus type 16 major capsid protein	Potato; tobacco	Biemelt et al. 2003; Warzecha et al. 2003
VP6 protein of rotavirus	Potato	Matsumura et al. 2002
Respiratory syncytial virus F and G-protein	Tobacco; tomato; apple	Belanger et al. 2000; Sandhu et al. 2000
*Enterotoxigenic *E. coli*	Potato; corn	Chikwamba et al. 2002; Mason et al. 1998; Tacket et al. 1998; Lauterslager et al. 2001
Enteropathogenic *E. coli*	Tobacco	Vieira da Silva et al. 2002
Vibrio cholerae toxin	Tobacco; tomato; potato	Daniell et al. 2001a; Jani et al. 2002; Arakawa et al. 1997, 1998, 1999
Tuberculosis ESAT–6 antigen	*Arabidopsis thaliana*	Rigano et al. 2003
Anthrax protective antigen	Tobacco	Azhar et al. 2002
Taenia solium cysticercosis peptides	Carrots, papaya	Sciutto et al. 2002

From vaccine production to administration, transgenic plant-derived vaccines offer significant advantages over other vaccine development strategies. These include advantages in manufacturing, packaging, storage, transportation, and most important, advantages in the ease of administration and safety for the recipient. Plant-based vaccines would eliminate the concern about transmission of human pathogens, remove the need for needles and syringes, and reduce the need for trained medical personnel to administer the vaccine. The plant cell wall would potentially protect the antigen in the stomach and intestine. Depending on the formulation of the plant-derived vaccine, there may be reduced or no requirement for refrigerated storage. In the developing world, transgenic plant vaccines may also be produced locally near the population to be vaccinated, even if some low-technology processing of the plant is required, e.g., corn germ meal (Streatfield et al. 2003; Chikwamba et al. 2002) or dehydrated tomato powder (Walmsley et al. 2003; Sala et al. 2003). These savings in vaccine-related production, supplies, and labor reduce the cost of each dose of vaccine.

This chapter describes recent phase 1 studies of oral transgenic plant-derived active vaccines in humans. To date, human studies have involved transgenic plant-derived vaccines consisting of plant organs (leaves or fruit) or crude extracts (dry powder), formulations made by low-cost food processing technology. Reports of human studies of plant-derived monoclonal antibodies (plantibodies) for passive immunization are expected soon.

How Transgenic Plant Vaccines are Made

The first prototype plant-derived vaccines were constructed in tobacco plants because of the ease of transformation and regeneration of this plant. Edible plants, such as potato, tomato, lettuce, carrots, and corn, have joined tobacco as hosts for foreign genes. The foreign DNA can be transiently introduced by infection of susceptible plants with the recombinant virus encoding foreign DNA (Mason and Arntzen 1997) or stably expressed by integrating the foreign DNA into the plant nuclear genome or into the circular chloroplast genome (Sala et al. 2003).

Most oral, plant-derived vaccine candidates have used stably transformed plants most commonly achieved using the bacterium *Agrobacterium tumefaciens*. Within this soil organism resides a tumor-inducing (Ti) plasmid containing transfer DNA or T-DNA. The Ti plasmid encodes factors that move a portion of the T-DNA into a plant cell and integrate it stably into the plant nuclear genome (Zambryski 1988). Foreign genes can be introduced into the T-DNA and transferred to the plant nucleus and randomly inserted into the chromosomal DNA at one or more sites. The promoter for the gene of interest can be either constitutive or tissue-specific, which affects the timing or the location of expression within the plant. The codon usage of the foreign gene of interest can be optimized for plant expression.

To transform plants using *A. tumefaciens*, cut surfaces of plant tissues are inoculated with bacteria containing the foreign gene in the Ti plasmid. The bacteria attach to plant cells at the wound site. The resulting leaf pieces are cultured on nutrient agar medium along with an antibiotic to kill the *A. tumefaciens*. A single transformed plant cell can produce a shoot that can be transplanted to soil and grown in a greenhouse or growth chamber. Because insertion of T-DNA may inhibit growth, many transgenic lines are screened to identify a transformant that expresses the foreign gene at high levels and does not adversely affect the plant.

Processed preparations derived from transgenic plants offer significant advantages over the whole fruit or vegetable. The processing must involve low heat and pressure so as not to denature the antigen (Streatfield et al. 2002). Examples include corn germ meal, corn flakes, dehydrated tomato powder, and banana flakes. These formulations are suitable for oral delivery and a large amount of antigen can be contained in a small volume of material and be easily ingested.

Human Studies: Transgenic Plant-Derived Vaccines

A large number of genes from human pathogens have been expressed in plants (Table 1). A few prototypical vaccines for human pathogens have been developed and tested in humans, including vaccines against enterotoxigenic *Escherichia coli*, norovirus, and hepatitis B.

Enterotoxigenic E. coli

Enterotoxigenic *E. coli* (ETEC) is responsible for diarrhea with dehydration and death in young children in the developing world and is one of the most common causes of traveler's diarrhea. ETEC includes a family of *E. coli* organisms that differ by O:H serogroup, fimbrial colonization factor antigens, and toxin type (heat labile toxin, or LT, and heat-stable toxin). A number of experimental vaccines against ETEC have been devised and some have undergone testing in humans with some success (Savarino et al. 2002). A vaccine must include a broad spectrum of antigens to be effective against diverse ETEC strains. Plant-derived antigens, inexpensive and possibly produced locally, could provide an efficient source of the multiple components for the ETEC vaccine of the future.

The initial focus in plant-derived ETEC vaccine development has been on the highly immunogenic LT of *E. coli*. This toxin consists of an enzymatically active A subunit with ADP ribosyl transferase activity associated with five immunogenic binding, or B, subunits, designated LT-B, which bind to the GM1 ganglioside present on epithelial cells. Antibody to LT-B could prevent binding of the toxin to the epithelial, thereby protecting against diarrhea. Immune responses to LT-B may offer short-term protection against infection with LT-producing *E. coli* (Clemens et al. 1988).

Transgenic Potatoes Expressing LT-B

Haq et al. transferred the gene encoding LT-B into tobacco and potato plants via *A. fumefaciens* and fed these tobacco leaves and potatoes to mice (Haq et al. 1995). The plant-derived LT-B assembled into pentameric structures and bound to ganglioside, like bacteria-derived LT-B. Each 5-g dose of transgenic potato tuber delivered 15–20 μg of rLT-B. Mice that consumed these potatoes developed serum IgG and mucosal IgA anti-LT-B, and some responses were similar to those of animals immunized with 20-μg doses of purified LT-B expressed in bacteria and given by oral gavage (Haq et al. 1995). In a subsequent study, mice fed three weekly doses of 5 g of tuber tissue containing either 20 or 50 μg of LT-B had higher levels of serum and mucosal anti-LT-B than those gavaged with 5 μg of bacterial LT-B (Mason et al. 1998). To prove that antibodies stimulated by the plant-derived vaccine were protective, vaccinated mice were challenged with 25 μg of LT. Although none of the vaccinated mice was completely protected, the potato vaccine provided a significant reduction in fluid accumulation in the patent mouse assay. Control mice fed nontransformed potatoes developed no antibodies and no reduction in secretion in the patent mouse assay (Mason et al. 1998).

These encouraging animal studies led to a phase 1 human study of the potato-derived vaccine (Tacket et al. 1998). Raw transgenic potatoes and control wild-type potatoes were peeled immediately before ingestion to remove the skin containing solanine, an alkaloid present in all raw potatoes, which can cause gastrointestinal upset. Potatoes were cut into small pieces and each dose weighed. Fourteen healthy adult volunteers ingested three oral doses of either 100 g of transgenic potato expressing LT-B (n=6), 50 g of transgenic potato (n=5), or 50 g of wild-type potato (n=3). Each dose of potato contained approximately 0.4–1.1 mg LT-B. The raw potatoes were generally well tolerated.

All volunteers who ingested transgenic potatoes developed circulating antibody secreting cells (ASCs) specific for LT after immunization (Tacket et al. 1998). These cells, detected in the peripheral blood approximately 7–10 days after immunization, reflect priming of the intestinal mucosal immune system. Similarly, ten (91%) of 11 volunteers who ingested transgenic potatoes developed at least fourfold rises in serum IgG anti-LT after immunization, and eight (73%) of 11 volunteers developed LT neutralizing antibody, indicating that the antibodies elicited by the potato vaccine were fully functional. Half the volunteers who ingested transgenic potatoes developed at least fourfold rises in sIgA in their stools.

In these studies, an oral transgenic potato-derived vaccine was successfully immunogenic even without co-administration of buffer or encapsulation to protect the antigen. This represented the first proof of the principle that transgenic plant vaccines formulated as whole vegetable could be immunogenic in humans. The door was opened for further refinement and development of other vaccines containing a cloned protective antigen.

Transgenic Corn Expressing LT-B

Vaccine antigens have also been successfully expressed in transgenic corn (Chikwamba et al. 2002; Streatfield et al. 2001). Corn-derived antigen is inexpensive to produce and can be scaled up rapidly, with corn generation time of 3–4 months. Corn-derived proteins can be expressed and concentrated at high levels of up to 10 mg/g in the corn germ, and the expressed protein is very similar to native protein (Hood et al. 1997; Woodard et al. 2003). Antigens expressed in transgenic corn are stable (Streatfield et al. 2002; Lamphear et al. 2002), and the product of the cloned gene is highly concentrated and homogeneous in corn germ (Streatfield et al. 2003; Lamphear et al. 2002).

Streatfield et al. developed a prototype transgenic corn containing the gene encoding LT-B (Streatfield et al. 2002). The transgenic corn vaccine was formulated as defatted corn germ meal, prepared by standard commercial processing techniques. Removal of fat concentrated the LT-B in the germ and also decreased the risk of the corn material becoming rancid with storage. The grinding of the defatted germ produced uniform particle size and a homogeneous distribution of LT-B.

In preclinical studies in mice, the transgenic corn germ meal was well tolerated and stimulated serum IgG responses and fecal IgA responses (Lamphear et al. 2002). Mice fed LT-B corn, but not control corn, were protected from intestinal secretion in the patent mouse assay (Streatfield et al. 2001). The degree of protection elicited by the LT-B corn was similar to that elicited by an equivalent amount of purified bacterial LT-B.

A clinical study of the safety and immunogenicity of the transgenic corn germ meal was conducted in which transgenic corn expressing 1 mg of LT-B of *E. coli* without buffer was fed to adult volunteers in three doses, each consisting of 2.1 g of plant material (Tacket et al. 2004). Seven (78%) of nine vaccinees developed at least fourfold rises in serum IgG anti-LT after vaccination, usually after the second or third dose of transgenic corn germ meal vaccine. The IgG titer peaked at day 56. Four (44%) of nine developed fourfold rises in serum IgA anti-LT; the IgA titer peaked at day 14. Seven (78%) of nine vaccinees developed specific IgA ASC. Four (44%) of nine vaccinees developed at least fourfold rises in stool sIgA anti-LT concentrations after vaccination (mean peak fold rise of 7.6 among responders).

Norovirus

Noroviruses are members of the *Caliciviridae* family which are the major cause of epidemic gastroenteritis in the United States (Glass et al. 2000). Included in this group is Norwalk virus (NV), which is of particular epidemiologic significance (Green et al. 1993; Deneen et al. 2000).

The major capsid protein of NV, cloned and expressed in insect cells, folds spontaneously into virus-like particles (VLPs) that lack nucleic acid (Xi et al. 1990). Norwalk VLPs administered with buffer are immunogenic when given orally to volunteers and are a potential vaccine candidate (Ball et al. 1999; Tacket et al. 2003).

When expressed in tobacco and potatoes, and the plant-derived recombinant NV particles are identical to those derived from insect cell culture. When transgenic potatoes were fed to mice, the plant-derived VLP stimulated serum and fecal antibody responses (Mason et al. 1996). A human study was conducted in which 24 healthy adult volunteers were randomized in a double-blind manner to receive one of three different regimens: (a) three doses of transgenic potato on days 0, 7, and 21 (n=10); (b) two doses of transgenic potato on days 0 and 21 and a dose of wild-type potato on day 7 (n=10); or (c) three doses of wild-type potato on days 0, 7, and 21 (n=4) (Tacket et al. 2000b). On the morning of dosing, transgenic and wild-type potatoes were peeled and cut into uniform 1-cm square cubes. Doses of 150 g of raw potato cubes were weighed on a scale and immediately ingested. Each dose contained approximately 500 μg of Norwalk virus capsid protein, half of which was assembled into virus-like particles.

Nineteen (95%) of 20 volunteers who ingested two or three doses of transgenic potatoes, and none who ingested wild-type potatoes, developed significant rises in the numbers of IgA ASC (range 6–280/10^6 peripheral blood mononuclear cells, PBMC). Thirteen of the 19 IgA ASC responses occurred after the first dose of transgenic potato. Four (20%) of 20 volunteers developed serum IgG anti-NVCP (mean 12-fold rise), and four (20%) of 20 volunteers (three of whom did not develop IgG responses) developed serum IgM anti-NVCP (mean sevenfold rise) after ingesting transgenic potatoes. Stool IgA anti-NVCP was detected in six (30%) volunteers who ingested transgenic potatoes (mean fold rise in titer of 17 among responders).

Hepatitis B Virus

Hepatitis B virus is prevalent worldwide and causes chronic hepatitis, cirrhosis, and hepatocellular carcinoma after parenteral or sexual transmission. The hepatitis B surface antigen (HBsAg) elicits protective immunity after intramuscular injection of three doses. An estimated 2 billion people are infected with hepatitis B virus, infections that could be prevented by vaccination. The parenteral vaccine is available and is widely used in the developed world, and the Global Advisory Group of the Expanded Program on Immunization and World Health Assembly have recommended that countries with higher than a 2% prevalence of HBV carriers add hepatitis B vaccine to their routine infant immunization schedules. However, the high price of parenteral vaccine has prevented the recommended vaccinations in some countries (Beutels 1998). An inexpensive, needle-less, plant-derived hepatitis B vaccine would therefore be a desirable public health tool for the control of hepatitis B.

Mason et al. first transformed tobacco plants with the gene encoding hepatitis B surface antigen (Mason et al. 1992). The recombinant plant-derived antigen formed spherical 22-nm particles identical to human serum-derived HBsAg. Subsequently, mice were immunized with the tobacco-derived HBsAg (Thanavala et al. 1995). The serum antibody and T cell immune responses in mice fed transgenic tobacco leaves were similar to those in mice immunized with commercial hepatitis B vaccine derived from yeast. This group then developed a transgenic potato line PAT-HB-7 expressing 1.1 μg of HBsAg per g of potato (Richter et al. 2000). When given to mice in three weekly doses of 5.5 μg HBsAg per dose along with 10 μg of cholera toxin (CT), a known mucosal adjuvant, this vaccine elicited a primary immune response measured by increases in specific serum antibody to a peak of 73 mIU/ml (Richter et al. 2000). These responses were boosted to a level of 1,679 mIU/ml by a single small dose of 0.5 μg of commercial hepatitis B vaccine delivered intraperitoneally. The plant-derived vaccine, delivering a small dose of antigen, had primed the animal for the unusually robust booster response to the intraperitoneal vaccine.

In subsequent studies, these investigators compared the immunogenicity of oral yeast-derived HBsAg and oral potato-derived HBsAg in mice (Kong et al. 2001). Yeast-derived HBsAg given as two doses of 150 μg each with bicarbonate buffer plus 10 μg of CT adjuvant did not stimulate serum antibody in mice. HBsAg given as three doses of 142 μg/dose delivered in 5 g of potatoes along with 10 μg of CT resulted in a peak of 103 mIU/ml of serum antibody after the third dose. (The protective level of serum antibody is 10 mIU/ml.) On electron microscopy, the plant-derived HBsAg had accumulated intracellularly, suggesting that a natural bioencapsulation of the antigen might protect it from degradation in the intestinal tract, while purified yeast-derived HBsg is not protected. A human study of this vaccine is not yet published.

Another group led by Kapusta introduced the HBsAg gene into lupin and lettuce plants (Kapusta et al. 1999). Mice fed transgenic lupin callus developed hepatitis B-specific antibodies. Three humans were fed two doses of transgenic lettuce containing 200 g and 150 g of lettuce leaves within 2 months (Kapusta et al. 1999). The amount of HBsAg in the lettuce varied from 0.1 to 0.5 μg/100 g, so the volunteers received approximately 0.2–1 μg of antibody in the 200-g dose. (For comparison, the licensed injectable hepatitis B vaccine contains 10 μg of HBsAg per adult dose.) After the second dose, all three volunteers developed anti-HBsAg antibody; two of the three had titers greater than 10 mIU/l, the minimum protective level of antibody against hepatitis B virus.

In a subsequent study, seven seronegative volunteers were immunized three times with fresh transgenic lettuce leaves on a 0–1-5-week schedule (Kapusta et al. 2001). The amount of HBsAg ranged from 0.51 to 0.94 μg per dose. Three weeks after the second immunization, antibody was detected in all seven volunteers, but not in control volunteers who received untransformed lettuce. The antibody responses were short-lived, but the third dose restimulated a rise in specific antibodies. Two weeks after the third dose, all volunteers had an increased level of specific antibody between 2 and 6.3 mIU/l, less than the protective level of 10 mIU/l, but still a significant rise from baseline.

Multivalent Transgenic Plant-Derived Diarrheal Disease Vaccine

Combination vaccines have been developed so that children can receive multiple vaccinations with a single injection and single encounter with a health care provider (Lagos et al. 1998). One of the advantages of plant-based vaccines is that plants that produce two or more antigens from different pathogens can be constructed. A prototype multicomponent vaccine was constructed in which cholera toxin B and A2 genes were fused to rotavirus enterotoxin and ETEC fimbrial genes and expressed in potato (Yu and Langridge 2001). When this vaccine was given to mice, serum and intestinal antibodies were detected. Such a multivalent vaccine might also include a transgenic plant-produced nontoxic derivative of LT that is a potent mucosal adjuvant when co-administered with another antigen (Douce et al. 1999).

Oral Tolerance

It is remarkable that the plant-derived vaccine protein is recognized within the context of the food delivery system, is processed as an antigen, and elicits an immune response. Immune tolerance is the usual result when the mucosal immune system encounters antigen (Strober et al. 1998). One potential safety concern about presentation of vaccine antigens in the context of food is that oral tolerance could be stimulated against the antigen. Theoretically, this could result in a suboptimal immune response if the individual were confronted with that antigen in the future during natural infection. Preliminary data in humans suggest that oral ingestion of multiple doses of KLH antigen actually primes serum and mucosal antibody responses to subsequent parenteral immunization with KLH, although T cell responses were inhibited (Husby et al. 1994). The safety and efficacy of the currently licensed oral vaccines offer reassurance that antigens can be delivered orally without induction of tolerance.

Regulatory Issues

Plant-derived vaccines should be produced, processed, and regulated as pharmaceutical biologic products (Stein and Webber 2001). The consistency and potency of the dose must be demonstrated; this may best be achieved by formulating the transgenic plant vaccine as a dehydrated powder or juice homogenate. Environmental concerns about mixing genetically modified pollen with other crops or weeds have been raised (Stokstad and Vogel 2003). This objection may be partially addressed by engineering the foreign gene into the chloroplast DNA (Ruf et al. 2001). Growing pharmaceutical crops in greenhouses or on small parcels of land will also prevent spread of genetically modified pollen. As higher levels of genetic expression are achieved through technical advances, land requirements for plant-derived vaccine production will decrease.

In the United States, the Animal and Plant Health Inspection Service (APHIS) of the USDA oversees the movement of plants between states and their release into the environment. A permit from APHIS is required to grow engineered plants that express a biologic drug in the field. Plants grown in an enclosed building such as a greenhouse or laboratory are considered contained if there are measures in place to prevent spread of pollen or seeds outside the facility. Transgenic plants must also be contained during transport. Transgenic food plants must not enter the human food supply, and the transgenic plant material must be strictly separated and clearly labeled.

Conclusion

Use of plants with medicinal properties is the traditional foundation of pharmaceutical medicine. Molecular genetic techniques allow the specific manipulation of ordinary food plants to produce drugs and biologics. In early phase 1 studies, prototype transgenic plant vaccines have been well tolerated and immunogenic. New formulations of these vaccines, such as corn germ meal and dehydrated tomato powder, are being developed and the regulatory framework for evaluating these biological products is adapting to the issues peculiar to this technology. The important developments in the future will be improving the immunogenicity of transgenic plant vaccines by delivery of higher amounts of antigen by improved expression or concentration of antigen; the co-administration of antigen with a mucosal adjuvant; and introduction of transgenic plants expressing multiple protective antigens of different pathogens. Transgenic plant technology may be a significant step toward the goal of developing less expensive childhood vaccines as well as inexpensive vaccines against emerging diseases.

Acknowledgements The author acknowledges the contribution of the staff of the Adult Clinical Studies Section and the Applied Immunology Section of the Center for Vaccine Development, University of Maryland, and the support of contract NO1-AI-65299, the Enteric Pathogens Research Unit, from NIAID/NIH and the University of Maryland General Clinical Research Center grant M01 RR165001, General Clinical Research Centers Program, National Center for Research Resources (NCRR), NIH.

References

Arakawa T, Chong DK, Merritt JL, Langridge WH (1997) Expression of cholera toxin B subunit oligomers in transgenic potato plants. Transgenic Res 6:403–413

Arakawa T, Chong DK, Langridge WH (1998) Efficacy of a food plant-based oral cholera toxin B subunit vaccine. Nat Biotechnol 16:292–297

Arakawa T, Yu J, Langridge WH (1999) Food plant-delivered cholera toxin B subunit for vaccination and immunotolerization. Adv Exp Med Biol 464:161–178

Azhar AM, Singh S, Anand KP, Bhatnagar R (2002) Expression of protective antigen in transgenic plants: a step towards edible vaccine against anthrax. Biochem Biophys Res Commun 299:345–351

Ball JM, Graham DY, Opekun AR, Gilger MA, Guerrero RA, Estes MK (1999) Recombinant Norwalk virus-like particles given orally to volunteers: phase I study. Gastroenterology 117:40–48

Belanger H, Fleysh N, Cox S, et al (2000) Human respiratory syncytial virus vaccine antigen produced in plants. FASEB J 14:2323–2328

Beutels P (1998) Economic evaluations applied to HB vaccination: general observations. Vaccine 16 [Suppl:]S84–S92

Biemelt S, Sonnewald U, Galmbacher P, Willmitzer L, Muller M (2003) Production of human papillomavirus type 16 virus-like particles in transgenic plants. J Virol 77:9211–9220

Chargelegue D, Vine ND, van Dolleweerd CJ, Drake PM, Ma JK (2000) A murine monoclonal antibody produced in transgenic plants with plant-specific glycans is not immunogenic in mice. Transgenic Res 9:187–194

Chikwamba R, Cunnick J, Hathaway D, McMurray J, Mason H, Wang K (2002) A functional antigen in a practical crop: LT-B producing maize protects mice against *Escherichia coli* heat labile enterotoxin (LT) and cholera toxin (CT). Transgenic Res 11:479–493

Clemens JD, Sack DA, Harris JR, et al (1988) Cross-protection by B subunit-whole cell cholera vaccine against diarrhea associated with heat-labile toxin-producing enterotoxigenic *Escherichia coli*: results of a large-scale field trial. J Infect Dis 158:372–377

Daniell H, Lee SB, Panchal T, Wiebe PO (2001a) Expression of the native cholera toxin B subunit gene and assembly as functional oligomers in transgenic tobacco chloroplasts. J Mol Biol 311:1001–1009

Daniell H, Streatfield SJ, Wycoff K (2001b) Medical molecular farming: production of antibodies, biopharmaceuticals and edible vaccines in plants. Trends Plant Sci 6:219–226

Deneen VC, Hunt JM, Paule CR, et al (2000) The impact of foodborne calicivirus disease: the Minnesota experience. J Infect Dis 181 [Suppl 2]:S281–S283

Devico AL, Fouts TR, Shata MT, Kamin-Lewis R, Lewis GK, Hone DM (2002) Development of an oral prime-boost strategy to elicit broadly neutralizing antibodies against HIV-1. Vaccine 20:1968–1974

DiPetrillo MD, Tibbetts T, Kleanthous H, Killeen KP, Hohmann EL (1999) Safety and immunogenicity of phoP/phoQ-deleted *Salmonella typhi* expressing *Helicobacter pylori* urease in adult volunteers. Vaccine 18:449–459

Douce G, Giannelli V, Pizza M, et al (1999) Genetically detoxified mutants of heat-labile toxin from *Escherichia coli* are able to act as oral adjuvants. Infect Immun 67:4400–4406

Edelman R (1997) Adjuvants for the future. In: Levine MM, Woodrow GC, Kaper JB, Cobon GS (eds) New generation vaccines. Marcel Dekker, New York, pp 173–192

Fischer R, Hoffmann K, Schillberg S, Emans N (2000) Antibody production by molecular farming in plants. J Biol Regul Homeost Agents 14:83–92

Gao Y, Ma Y, Li M, et al (2003) Oral immunization of animals with transgenic cherry tomatillo expressing HBsAg. World J Gastroenterol 9:996–1002

Giddings G (2001) Transgenic plants as protein factories. Curr Opin Biotechnol 12:450–454

Giddings G, Allison G, Brooks D, Carter A (2000) Transgenic plants as factories for biopharmaceuticals. Nat Biotechnol 18:1151–1155

Glass RI, Noel J, Ando T, et al (2000) The epidemiology of enteric caliciviruses from humans: a reassessment using new diagnostics. J Infect Dis 181 [Suppl 2]:S254–S261

Grangette C, Muller-Alouf H, Geoffroy M, Goudercourt D, Turneer M, Mercenier A (2002) Protection against tetanus toxin after intragastric administration of two recombinant lactic acid bacteria: impact of strain viability and in vivo persistence. Vaccine 20:3304–3309

Green KY, Lew JF, Jiang X, Kapikian AZ, Estes MK (1993) Comparison of the reactivities of baculovirus-expressed recombinant Norwalk virus capsid antigen with those of the native Norwalk virus antigen in serologic assays and some epidemiologic observations. J Clin Microbiol 31:2185–2191

Haq TA, Mason HS, Clements JD, Arntzen CJ (1995) Oral immunization with a recombinant bacterial antigen produced in transgenic plants. Science 268:714–716

Hood EE, Witcher DR, Maddock S, Meyer T, Baszczynski C, Bailey M, Flynn P, Register J, Marshall L, Bond D, Kulisek E, Kusnadi A, Evangelista R, Nikolov Z, Wooge C, Mehigh RJ, Hernan R, Kappel WK, Ritland D, Li CP, Howard JA (1997) Commercial production of avidin from transgenic maize: characterization of transformant, production, processing, extraction and purification. Mol Breed 3:291–306

Hood EE, Kusnadi A, Nikolov Z, Howard JA (1999) Molecular farming of industrial proteins from transgenic maize. Adv Exp Med Biol 464:127–147

Huang Z, Dry I, Webster D, Strugnell R, Wesselingh S (2001) Plant-derived measles virus hemagglutinin protein induces neutralizing antibodies in mice. Vaccine 19:2163–2171

Husby S, Mestecky J, Moldoveanu Z, Holland S, Elson CO (1994) Oral tolerance in humans. T cell but not B cell tolerance after antigen feeding. J Immunol 152:4663–4670

Jani D, Meena LS, Rizwan-ul-Haq QM, Singh Y, Sharma AK, Tyagi AK (2002) Expression of cholera toxin B subunit in transgenic tomato plants. Transgenic Res 11:447–454

Kapusta J, Modelska A, Figlerowicz M, et al (1999) A plant-derived edible vaccine against hepatitis B virus. FASEB J 13:1796–1799

Kapusta J, Modelsak A, Pniewski T, et al (2001) Oral immunization of human with transgenic lettuce expressing hepatitis B surface antigen. Adv Exp Med Biol 495:299–303

Kong Q, Richter L, Yang YF, Arntzen CJ, Mason HS, Thanavala Y (2001) Oral immunization with hepatitis B surface antigen expressed in transgenic plants. Proc Natl Acad Sci U S A 98:11539–11544

Kusnadi AR, Evangelista RL, Hood EE, Howard JA, Nikolov ZL (1998a) Processing of transgenic corn seed and its effect on the recovery of recombinant beta-glucuronidase. Biotechnol Bioeng 60:44–52
Kusnadi AR, Hood EE, Witcher DR, Howard JA, Nikolov ZL (1998b) Production and purification of two recombinant proteins from transgenic corn. Biotechnol Prog 14:149–155
Lagos R, Kotloff K, Hoffenbach A, et al (1998) Clinical acceptability and immunogenicity of a pentavalent parenteral combination vaccine containing diphtheria, tetanus, acellular pertussis, inactivated poliomyelitis and *Haemophilus influenzae* type b conjugate antigens in two-, four- and six-month-old Chilean infants. Pediatr Infect Dis J 17:294–304
Lamphear BJ, Streatfield SJ, Jilka JM, et al (2002) Delivery of subunit vaccines in maize seed. J Control Release 85:169–180
Larrick JW, Thomas DW (2001) Producing proteins in transgenic plants and animals. Curr Opin Biotechnol 12:411–418
Lauterslager TG, Florack DE, van der Wal TJ, et al (2001) Oral immunisation of naive and primed animals with transgenic potato tubers expressing LT-B. Vaccine 19:2749–2755
Ma JK, Hiatt A, Hein M, et al (1995) Generation and assembly of secretory antibodies in plants. Science 268:716–719
Ma SW, Zhao DL, Yin ZQ, et al (1997) Transgenic plants expressing autoantigens fed to mice to induce oral immune tolerance. Nat Med 3:793–796
Ma JK, Hikmat BY, Wycoff K, et al (1998) Characterization of a recombinant plant monoclonal secretory antibody and preventive immunotherapy in humans. Nat Med 4:601–606
Ma Y, Lin SQ, Gao Y, et al (2003) Expression of ORF2 partial gene of hepatitis E virus in tomatoes and immunoactivity of expression products. World J Gastroenterol 9:2211–2215
Marquet-Blouin E, Bouche FB, Steinmetz A, Muller CP (2003) Neutralizing immunogenicity of transgenic carrot (*Daucus carota* L)-derived measles virus hemagglutinin. Plant Mol Biol 51:459–469
Mason HS, Arntzen CJ (1995) Transgenic plants as vaccine production systems. Trends Biotechnol 13:388–392
Mason HS, Lam DM, Arntzen CJ (1992) Expression of hepatitis B surface antigen in transgenic plants. Proc Natl Acad Sci U S A 89:11745–11749
Mason HS, Ball JM, Shi JJ, Jiang X, Estes MK, Arntzen CJ (1996) Expression of Norwalk virus capsid protein in transgenic tobacco and potato and its oral immunogenicity in mice. Proc Natl Acad Sci U S A 93:5335–5340
Mason HS, Haq TA, Clements JD, Arntzen CJ (1998) Edible vaccine protects mice against *Escherichia coli* heat-labile enterotoxin (LT): potatoes expressing a synthetic LT-B gene. Vaccine 16:1336–1343
Matsumura T, Itchoda N, Tsunemitsu H (2002) Production of immunogenic VP6 protein of bovine group A rotavirus in transgenic potato plants. Arch Virol 147:1263–1270
McGarvey PB, Hammond J, Dienelt MM, et al (1995) Expression of the rabies virus glycoprotein in transgenic tomatoes. Biotechnology (N Y) 13:1484–1487
Modelska A, Dietzschold B, Sleysh N, et al (1998) Immunization against rabies with plant-derived antigen. Proc Natl Acad Sci U S A 95:2481–2485
Nardelli-Haefliger D, Kraehenbuhl JP, Curtiss R III, et al (1996) Oral and rectal immunization of adult female volunteers with a recombinant attenuated *Salmonella typhi* vaccine strain. Infect Immun 64:5219–5224
Orr N, Galen JE, Levine MM (1999) Expression and immunogenicity of a mutant diphtheria toxin molecule CRM(197), and its fragments in *Salmonella typhi* vaccine strain CVD 908-htrA. Infect Immun 67:4290–4294
Pouwels PH, Leer RJ, Boersma WJ (1996) The potential of *Lactobacillus* as a carrier for oral immunization: development and preliminary characterization of vector systems for targeted delivery of antigens. J Biotechnol 44:183–192
Richter LJ, Thanavala Y, Arntzen CJ, Mason HS (2000) Production of hepatitis B surface antigen in transgenic plants for oral immunization. Nat Biotechnol 18:1167–1171

Rigano MM, Alvarez ML, Pinkhasov J, et al (2003) Production of a fusion protein consisting of the enterotoxigenic Escherichia coli heat-labile toxin B subunit and a tuberculosis antigen in *Arabidopsis thaliana*. Plant Cell Rep 22:502–508

Ruf S, Hermann M, Berger IJ, Carrer H, Bock R (2001) Stable genetic transformation of tomato plastids and expression of a foreign protein in fruit. Nat Biotechnol 19:870–875

Sala F, Manuela RM, Barbante A, Basso B, Walmsley AM, Castiglione S (2003) Vaccine antigen production in transgenic plants: strategies, gene constructs and perspectives. Vaccine 21:803–808

Sandhu JS, Krasnyanski SF, Domier LL, Korban SS, Osadjan MD, Buetow DE (2000) Oral immunization of mice with transgenic tomato fruit expressing respiratory syncytial virus-F protein induces a systemic immune response. Transgenic Res 9:127–135

Savarino SJ, Hall ER, Bassily S, et al (2002) Introductory evaluation of an oral, killed whole cell enterotoxigenic *Escherichia coli* plus cholera toxin B subunit vaccine in Egyptian infants. Pediatr Infect Dis J 21:322–330

Scheppler L, Vogel M, Zuercher AW, et al (2002) Recombinant *Lactobacillus johnsonii* as a mucosal vaccine delivery vehicle. Vaccine 20:2913–2920

Sciutto E, Fragoso G, Manoutcharian K, et al (2002) New approaches to improve a peptide vaccine against porcine *Taenia solium* cysticercosis. Arch Med Res 33:371–378

Smith G, Walmsley A, Polkinghorne I (1997) Plant-derived immunocontraceptive vaccines. Reprod Fertil Dev 9:85–89

Smith ML, Mason HS, Shuler ML (2002) Hepatitis B surface antigen (HBsAg) expression in plant cell culture: kinetics of antigen accumulation in batch culture and its intracellular form. Biotechnol Bioeng 80:812–822

Stein KE, Webber KO (2001) The regulation of biologic products derived from bioengineered plants. Curr Opin Biotechnol 12:308–311

Stokstad E, Vogel G (2003) Agrobiotechnology. Mixed message could prove costly for GM crops. Science 302:542–543

Streatfield SJ, Jilka JM, Hood EE, et al (2001) Plant-based vaccines: unique advantages. Vaccine 19:2742–2748

Streatfield SJ, Mayor JM, Barker DK, Brooks C, Lamphear BJ, Woodard SL, Beifuss KK, Vicuna DV, Massey LA, Horn ME, Delaney DD, Nikoov ZL, Hood EE, Jilka JM, Howard JA (2002) Development of an edible subunit vaccine in corn against enterotoxigenic strains of *Escherichia coli*. In Vitro Cell Dev Biol Plant 38:11–17

Streatfield SJ, Lane JR, Brooks CA, et al (2003) Corn as a production system for human and animal vaccines. Vaccine 21:812–815

Strober W, Kelsall B, Marth T (1998) Oral tolerance. J Clin Immunol 18:1–30

Tackaberry ES, Dudani AK, Prior F, et al (1999) Development of biopharmaceuticals in plant expression systems: cloning, expression and immunological reactivity of human cytomegalovirus glycoprotein B (UL55) in seeds of transgenic tobacco. Vaccine 17:3020–3029

Tackaberry ES, Prior F, Bell M, et al (2003) Increased yield of heterologous viral glycoprotein in the seeds of homozygous transgenic tobacco plants cultivated underground. Genome 46:521–526

Tacket CO, Kelly SM, Schodel F, et al (1997) Safety and immunogenicity in humans of an attenuated *Salmonella typhi* vaccine vector strain expressing plasmid-encoded hepatitis B antigens stabilized by the Asd-balanced lethal vector system. Infect Immun 65:3381–3385

Tacket CO, Mason HS, Losonsky G, Clements JD, Levine MM, Arntzen CJ (1998) Immunogenicity in humans of a recombinant bacterial antigen delivered in a transgenic potato. Nat Med 4:607–609

Tacket CO, Galen J, Sztein MB, et al (2000a) Safety and immune responses to attenuated *Salmonella enterica* serovar typhi oral live vector vaccines expressing tetanus toxin fragment C. Clin Immunol 97:146–153

Tacket CO, Mason HS, Losonsky G, Estes MK, Levine MM, Arntzen CJ (2000b) Human immune responses to a novel Norwalk virus vaccine delivered in transgenic potatoes. J Infect Dis 182:302–305

Tacket CO, Sztein MB, Losonsky GA, Wasserman SS, Estes MK (2003) Humoral, mucosal, and cellular immune responses to oral Norwalk virus-like particles in volunteers. Clin Immunol 108:241–247

Tacket CO, Pasetti MF, Edelman R, Howard JA, Streatfield S (2004) Immunogenicity of recombinant LT-B delivered orally to humans in transgenic corn. Vaccine 22:4385–4389

Thanavala Y, Yang YF, Lyons P, Mason HS, Arntzen C (1995) Immunogenicity of transgenic plant-derived hepatitis B surface antigen. Proc Natl Acad Sci U S A 92:3358–3361

Vieira da Silva J, Garcia AB, Vieira Flores VM, Sousa de Macedo Z, Medina-Acosta E (2002) Phytosecretion of enteropathogenic *Escherichia coli* pilin subunit A in transgenic tobacco and its suitability for early life vaccinology. Vaccine 20:2091–2101

Walmsley AM, Alvarez ML, Jin Y, et al (2003) Expression of the B subunit of *Escherichia coli* heat-labile enterotoxin as a fusion protein in transgenic tomato. Plant Cell Rep 21:1020–1026

Warzecha H, Mason HS, Lane C, et al (2003) Oral immunogenicity of human papillomavirus-like particles expressed in potato. J Virol 77:8702–8711

Webster DE, Cooney ML, Huang Z, et al (2002a) Successful boosting of a DNA measles immunization with an oral plant-derived measles virus vaccine. J Virol 76:7910–7912

Webster DE, Thomas MC, Strugnell RA, Dry IB, Wesselingh SL (2002b) Appetising solutions: an edible vaccine for measles. Med J Aust 176:434–437

Woodard SL, Mayor JM, Bailey MR, et al (2003) Maize (*Zea mays*)-derived bovine trypsin: characterization of the first large-scale, commercial protein product from transgenic plants. Biotechnol Appl Biochem 38:123–130

Xi JN, Graham DY, Wang KN, Estes MK (1990) Norwalk virus genome cloning and characterization. Science 250:1580–1583

Yu J, Langridge WH (2001) A plant-based multicomponent vaccine protects mice from enteric diseases. Nat Biotechnol 19:548–552

Yusibov V, Hooper DC, Spitsin SV, et al (2002) Expression in plants and immunogenicity of plant virus-based experimental rabies vaccine. Vaccine 20:3155–3164

Zambryski P (1988) Basic processes underlying *Agrobacterium*-mediated DNA transfer to plant cells. Annu Rev Genet 22:1–30

Zegers ND, Kluter E, van Der SH, et al (1999) Expression of the protective antigen of *Bacillus anthracis* by *Lactobacillus casei*: towards the development of an oral vaccine against anthrax. J Appl Microbiol 87:309–314

Zeitlin L, Olmsted SS, Moench TR, et al (1998) A humanized monoclonal antibody produced in transgenic plants for immunoprotection of the vagina against genital herpes. Nat Biotechnol 16:1361–1364

Zhang GG, Rodrigues L, Rovinski B, White KA (2002) Production of HIV-1 p24 protein in transgenic tobacco plants. Mol Biotechnol 20:131–136

Index